Gas Syringe Experiments

for School and College Chemistry

Gas Syringe Experiments

for School and College Chemistry

Martin Rogers, M.A.
Master of the Queen's Scholars,
Westminster School

Heinemann Educational Books Ltd
London

Heinemann Educational Books
London Edinburgh Melbourne Toronto Cape Town
Singapore Auckland Ibadan Hong Kong Nairobi

SBN 435 64770 9

First Published 1970

Published by Heinemann Educational Books Ltd
48 Charles Street, London W1X 8AH

Printed in Great Britain by
Fletcher & Son Ltd, Norwich

Preface

The 100 cm³ glass syringe was first used for experiments with gases as long ago as 1931 by Dr Max Schmidt of Hamburg. It is remarkable that over thirty years elapsed before this versatile piece of apparatus became known in this country. It might have remained unknown for longer if I had not been fortunate enough to be sent on a tour of European schools by Professor H. F. Halliwell, who was at that time Organiser of the Nuffield Foundation O-level Chemistry Project. During the tour I visited Dr Adolf of the Hamburg Institute of Education who introduced me to syringe experiments. I am most grateful to Dr Adolf and to Dr Böse, Dr Adolf's predecessor at the Institute, who is an expert on syringe experiments and who gave me valuable advice on techniques. Many of the experiments in this book were first performed in Hamburg. I have modified them to suit the apparatus available in this country and in some cases have been able to introduce improvements. Other experiments have been developed at Westminster by myself and my pupils; yet others by colleagues who have passed their ideas on to me. To acknowledge the authorship of all these experiments is a formidable task and I hope I will be forgiven if my references are not complete in this respect. I am particularly grateful also to W. J. Porterfield, A. J. Neuberger, N. J. Margerison, P. W. K. Rundell, N. S. Snobel and D. B. L. Gifford for trying out many of the experiments and for suggesting valuable improvements.

If this book succeeds in encouraging science masters and their pupils to perform more experiments with gases it will have achieved its objective.

I am anxious to collect further syringe experiments and to improve the techniques for the ones described in this book. I shall therefore welcome suggestions from anyone who tries out these experiments.

M. J. W. Rogers

Westminster School

Acknowledgements

Experiments with syringes are to be found in the German literature in

Der mathematische und naturwissenschaftliche Unterricht
Praktische Schulphysik, Schulchemie, Schulbiologie
Praxis der Naturwissenschaften
Zeitschrift für den physikalen und chemischen Unterricht

and in *Das Kolbenprobergerät* by Robert Böse and Max Schmidt.

In addition to those mentioned in the Introduction, I am indebted to Dr A. Kemper who drew my attention to several syringe experiments.

In England I have received help and suggestions from Mr D. M. Stebbens and Mr D. W. Muffett (experiments 4.6 and 5.2), Mr C. V. Platts (experiment 7.4) and Mr K. W. Badman (experiments 4.4 and 5.1). Experiments 1.1, 2.1, 2.8 and 4.2 are to be found in the Nuffield O-level Chemistry course.

List of Contents

Chapter Three
Organic Analysis

Chapter Four
Rates of Reaction and Equilibria

Chapter Five
The Molecular Weight of Gases and Volatile Liquids

Chapter Six
Allotropy

Chapter Seven
The Physical Properties of Gases

Index

Introduction

Why do experiments with gases?

In the past, experiments involving the measurement of volumes of gases have been neglected. This is largely because of the awkwardness of the technique and the limitation imposed by collecting gases over water. The upturned burette is cheap but messy; the use of mercury is prohibitively expensive. A few enthusiasts have persevered but on the whole quantitative experiments with gases have become rare at school level. The introduction of simple techniques with gas syringes opens up the whole field again, making a review of the importance of experiments with gases well worth while.

The first advantage of the gaseous phase is that in it substances have their simplest structural form. It is therefore a good starting-point for the consideration of the nature of the structure of matter. It is not by chance that the early work of Gay-Lussac, Dalton and Avogadro concerned gases. From the practical point of view Avogadro's hypothesis leads to the idea of 'counting out' molecules by volume. With solids and liquids we have to 'count out' atoms and molecules by weighing; that is to say by knowing the atomic weights and weighing out a given number of gram atoms. Once our pupils have grasped the significance of Avogadro's hypothesis we can use the measurement of gaseous volumes to determine formulae and work out equations from first principles. Measuring the volume of a gas with a syringe is a much simpler operation than weighing out a solid. A good example of the approach to formulae and equations via gases is given in the Nuffield O-level Chemistry Course, *The Basic Course*, topics 11 and 17. Within the structure of this approach there are of course a number of experiments which confirm Gay-Lussac's Law.

The determination of molecular weights is another area in which gases play an important role. Using Avogadro's hypothesis Cannizzaro was able first to show that hydrogen is diatomic and, secondly, to find approximate values for the molecular and hence atomic weights of gases. It is possible with lightweight plastic syringes to find the densities of gases and so, like Cannizzaro, to determine molecular weights. The extension of this idea to volatile liquids can also be covered by syringe experiments. These experiments use a much simpler technique than that of Dumas and Victor Meyer and give results which are quite satisfactory for school work. The fact that some of these experiments (notably those with nitrogen dioxide and acetic acid) give 'unexpected' results can be used as a lead into the ideas of dissociation, association and equilibria. In the case of equilibria gases offer a unique advantage; gaseous equilibria reach the point of equilibrium relatively quickly. The temperature can be varied and the relationship between 'k' the equilibrium constant and 't' the temperature investigated. Another means of determining molecular weight is that which depends on Graham's Law of Diffusion. In fact the method used depends on effusion rather than diffusion but the results are the same. This experiment, which requires only a syringe and a tube with a pinhole in it, can be treated in an investigational way.

There are many ways in which rates of reactions may be measured. If a reaction in which a gas is evolved is chosen a syringe may be used to determine the rate of evolution of the gas. The temperature and concentration of the reactants may then be varied and the effects noted. The effect of catalysts may also be studied.

Syringe experiments also fulfil a useful function in the field of organic chemistry. The formulae of the gaseous hydrocarbons can be found by burning a known volume of a gas and measuring the volume of carbon dioxide formed. Liquid hydrocarbons and related compounds are susceptible to rather more sophisticated techniques, but the results are accurate enough for the formulae to be determined.

The case for gases is strong. Given the technique, and a certain amount of practice, the teaching of chemistry can be greatly enriched by the experiments which are described in this book.

The care of syringes

The 100 cm³ glass syringe is a precision instrument in which both barrel and plunger have been ground to very close tolerances; it is therefore very important that the syringe should be kept clean. It is recommended that after each experiment or series of experiments the syringe should be washed as follows:

Remove the plunger from the barrel and wash first with water, then with alcohol and finally with acetone or ether. Alternatively the syringe may be washed with distilled water and allowed to dry, or washed with distilled water and alcohol and dried by sucking air through it with a pump. In the latter case ensure that the end is covered with a filter-paper to reduce the risk of drawing dust into the syringe while drying it.

Fig. 0.1A

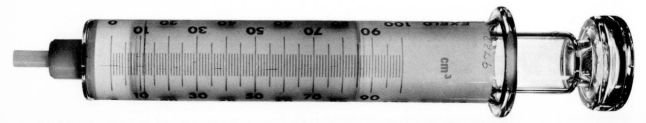

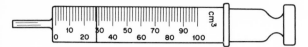

Fig. 0.1B

It is important to warn pupils against carrying the syringes in such a way that the plunger falls out. It is possible to guard against this by tying a piece of string round the neck of the plunger and round the end of the barrel in such a way that the plunger can pull out to the 100 cm³ mark but no further.

Glass syringes are best stored in the polystyrene packing in which they are despatched. Failing this cover the bottom of a drawer with cotton-wool to a depth of about 1 cm and lay the syringes on it.

50 cm³ plastic syringes are suitable for some of the experiments given in this book. They should be cleaned with water, but not with organic solvents as some of these attack the plastic plunger. Plastic syringes are not as smooth running as the glass syringes. To ensure that the gas in the syringe is at atmospheric pressure vibrate the end of the plunger up and down two or three times; it will then move to its right value.

Hints on using ancillary syringe apparatus

The following apparatus is available for use with 100 cm, glass syringes:

Fig. 0.2A

Fig. 0.2B

Transparent silica-glass tubing (Fig. 0.2). Ordinary hard glass is not strong enough for purposes of many of these experiments and transparent silica-glass tubing is recommended. Tubing should be about 8 mm outside diameter and 1 mm in wall thickness. For most experiments a tube of about 15 cm length is sufficient; occasionally a longer tube is necessary.

For most experiments it is important to have as little dead space as possible between the syringes, and it is therefore advisable to fill the silica tube with the substance which is placed in it, leaving as little air space as possible. In order to stop the material coming out of the end of the tubes it is advisable to insert small pieces of glass rod at either end, see Fig. 0.2A and 0.2B above. The glass rod should be of a diameter slightly smaller than the internal diameter of the silica tube. A tube with a 2 cm diameter blown in it is used in some experiments.

Fig. 0.3

Three-way stopcock (Fig. 0.3). The three-way stopcocks for use with syringe apparatus must have arms of a narrow bore, that is to say 1–1·5 mm internal diameter. It is convenient to have the outside diameter the same as that of the nozzle of the syringe or the silica-glass tubing, i.e. 7–8 mm.

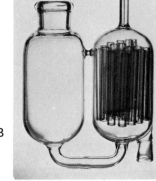

Fig. 0.4B

Fig. 0.4A

Orsat pipette (Fig. 0.4). For absorbing gases such as carbon dioxide an Orsat pipette is extremely useful. This type of pipette contains tubes which ensure a maximum surface area for contact between the liquid and the gas, see the diagram Fig. 0.4. The pipette may also be used as a manometer by marking the level of the liquid with a felt pen.

Absorbtion of gases using a diffuser.

A quicker way of absorbing gases is by means of a sintered glass diffuser. The gas is lead via a three-way stop-cock to a diffuser immersed in solution in a boiling tube or small flask. The gases emerge through a tube connected to a three-way stop-cock.

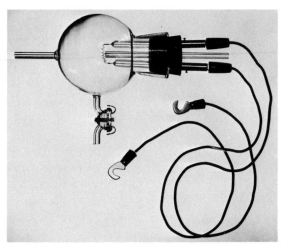

Fig. 0.5B

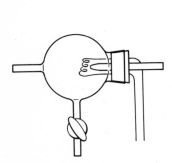

Fig. 0.5A

Combustion pipette (Fig. 0.5). A combustion pipette is needed for all experiments in which gases are burnt. The chamber, which must be made of a tough hard glass, contains a platinum spiral which is heated electrically. Take care to see that the spiral is immediately opposite the tube through which the gases enter the pipette and is not more than 0·2 cm from it.

Manometer (Fig. 0.6). A manometer is useful for experiments in which the syringes have to be greased. Under these conditions it is not possible to 'feel' the pressure of the gases and hence a manometer should be included. The liquid in the manometer should be a light machine oil. A three-way stopcock should be used in conjunction with the manometer so that it is not always connected with the syringes.

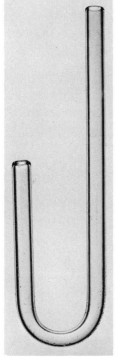

Fig. 0.6A

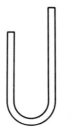

Fig. 0.6B

Syringe bench (Fig. 0.9, p. 8) ***and syringe clamps*** (Fig. 0.7). A syringe bench is available as are syringe clamps. The complete bench is made to hold all the apparatus required for almost every experiment in this book. It is possible, however, in many experiments to use syringe clamps with normal stands and bosses. Syringe clamps can be easily made with a piece of wood and two plastic-coated spring clips.

Fig. 0.7

Heating syringes

The syringe itself may be heated by immersion in hot water (or any other liquid). A thermostatically controlled water bath is ideal, but a beaker of water is quite adequate in most cases.

Two types of electrical heating are available; the syringe furnace and the syringe light bulb oven. The furnace has a maximum operating temperature of over 250 °C, but syringes tend to distort and leak at high temperatures. The syringe oven has a maximum temperature of about 120 °C and may be used for the determination of molecular weights (see p. 68 *ff*).

Heating the silica tube

This is normally done with a Bunsen burner, but for some experiments it has been found to be more effective to bind the silica tube with nickel chromium wire and to heat it electrically. A 'Variac', giving variable voltage supply provides a simple way of controlling the temperature. A suitable transformer may be used in conjunction with a variable resistance.

Units

In dealing with syringe experiments it is helpful to use a unit which may be unfamiliar to some chemists, the 'millimole'. Its definition as a thousandth of a mole is obvious but its advantages are less obvious until one considers the scale of working suitable for 100 cm³ syringes. A millimole of any gas at room temperature occupies about 24 cm³, about 22·4 cm³ at N.T.P. (What a pity that S.I. units do not include the millilitre.) Thus a standard syringe will hold up to 4 millimoles of a gas at room temperature. The weight of a millimole of a compound is its molecular weight in milligrammes. In some experiments it is only necessary to know how many millimoles of a gas are evolved to the nearest whole number. In such cases it is particularly useful to be aware of the simple volume relationship.

Precautions

Before starting an experiment with syringes the following precautions should be taken:

(1) See that the syringes are clean and smooth running.
(2) See that there is no unnecessary 'dead space' in the apparatus. The nozzle of the syringe, three-way stopcock, silica tube and other apparatus should fit as close together as possible.
(3) Check that there is no leak in the apparatus. This is best done by closing the apparatus to the air and pushing air from one syringe into the other and back again. If this occurs without loss of volume then the apparatus is reasonably gas-tight. Leakage usually occurs if the rubber connecting tubes are not sufficiently tight.
(4) When filling the syringe with a particular gas it is wise to flush the syringe out two or three times before finally putting a measured volume of the gas in it.

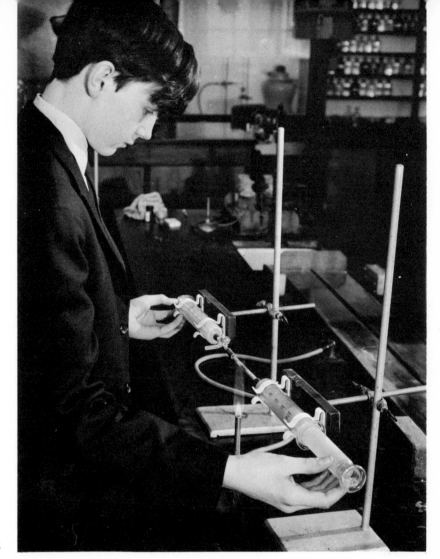

Fig. 0.8

(5) When passing gases backwards and forwards from one syringe to
another keep your hands on both syringes so that no excess
pressure builds up. This will minimise the risk of any leak.

**Availability of syringe
apparatus**

Syringes and ancillary apparatus are made in England by W. G. Flaig
& Sons Ltd., Exelo Works, Margate Road, Broadstairs, Kent.

The company also manufactures the gas syringe recorder which is
referred to in Chapter Four.

The syringes and apparatus may be obtained directly through
W. G. Flaig & Sons Ltd. or through suppliers of laboratory equipment.

A cheaper glass syringe, suitable for less accurate work, is also
available.

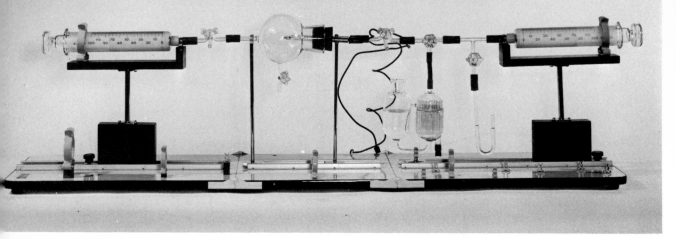

Fig. 0.9 Syringe bench with two syringes, combustion pipette, Orsat pipette and manometer

Film

A film on syringe experiments is available from Creative Services Division, Esso Petroleum. It is one of the Esso Science Teaching Films and is entitled *Gas Syringes*. In it the author discusses the care of syringes and demonstrates five of the experiments described in this book.

Chapter One

Introductory Experiments with Syringes

1.1   Finding the percentage of oxygen in the air

Introduction

This is a good experiment with which to introduce syringe techniques. It belongs in a chemistry course to the examination of the properties of the air. Having shown that there is an 'active part' to the air, and having found that certain metals gain weight on heating in air, this experiment may be introduced as a means of finding out how much of the 'active' gas or oxygen is present. It is suitable for use both as a demonstration experiment, with the pupils joining in with the weighings and other measurements, and as a pupil experiment. There is no reason why boys and girls should not perform experiments with glass syringes themselves provided they are carefully trained in their use, but some teachers may prefer to keep the more expensive glass syringes to themselves and allow their pupils to use only plastic syringes or the cheaper glass variety. Reasonable results can be obtained from this experiment when plastic syringes are used. The scope of the experiment may be extended by weighing the silica tube and copper before and after the experiment. The gain in weight is equal to the weight of oxygen which has reacted with the copper. As the volume of oxygen is known the density of the gas can be calculated.

Apparatus

2   100 cm³ glass syringes
1   three-way stopcock
1   silica-glass tube, with two 2 cm pieces of glass rod to fit inside
2   syringe clamps, with bosses and stands, or syringe bench

Simplified apparatus for pupil use:

2   50 cm³ plastic syringes
1   silica-glass tube as above
2   syringe clamps, bosses and stands

Chemicals
Wire-form copper (II) oxide, laboratory reagent quality
Source of hydrogen

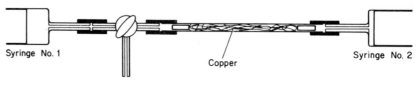

Syringe No. 1

Copper

Syringe No. 2

Fig. 1.1

Procedure

Preparation of wire-form copper powder:

A very suitable form of copper with a large surface area may be prepared as follows. Three-quarters fill the silica tube with wire-form copper oxide. Connect the tube to the source of hydrogen (preferably a cylinder) and pass a steady stream of the gas through the copper oxide in the tube. After testing to ensure that no air remains in the tube heat it until the copper oxide is completely reduced to copper. Allow the copper to cool down in a stream of hydrogen before disconnecting. It is worth preparing a quantity of the copper on a larger scale and keeping it for future use.

Assemble the apparatus as shown in Fig. 1.1 and test it to see that it is gas-tight (see p. 6). The silica tube should be filled with 'wire-form' copper except for the ends which are filled with the two glass rods. This cuts down dead space between the syringes to a minimum and prevents the copper from entering the syringes. Draw in 100 cm³ of air through the three-way stopcock into syringe No. 1. Turn the stopcock so that syringes Nos. 1 and 2 are connected to each other but not to the atmosphere. Heat the silica tube with a medium Bunsen flame. Pass the air backwards and forwards through the heated tube two or three times until the volume of the gas ceases to contract. The copper will oxidise to black copper oxide. Cool the tube down with a damp cloth (silica glass will not crack) and measure the volume of the remaining gas in syringe No. 1 again. Make sure that the gas is at atmospheric pressure by rotating the plunger gently. It is possible to 'feel' the pressure and thus avoid the use of a manometer.

Procedure with simplified apparatus:

The procedure is similar except that 50 cm³ of air must be drawn into one syringe before the apparatus is assembled. Plastic syringes must be lubricated. As they run less smoothly than glass syringes it is more difficult to 'feel' atmospheric pressure.

Results

With fresh air (see the next experiment for a discussion on class-room air) 100 cm³ of air will contract to 79·5 or 80 cm³ after the experiment, giving a result of 20·5 or 20 cm³ of oxygen. Errors of more than 1% should not be expected with glass syringes of the better type. Plastic syringes give results plus or minus 5%.

1.2   Determining the percentage of oxygen in exhaled air

Introduction

This experiment follows naturally from experiment 1.1. The determination of oxygen and carbon dioxide (experiment 1.3) in exhaled air can be performed as one experiment using the apparatus for this experiment with the Orsat pipette for experiment 1.3. The experiment is suitable for pupils or as a demonstration by the teacher.

Apparatus

2   100 cm³ glass syringes
1   three-way stopcock
1   silica-glass tube, with two 2 cm pieces of glass rod to fit inside
2   syringe clamps, with bosses and stands, or a syringe bench

Simplified apparatus for pupil use:

2   50 cm³ plastic syringes
1   silica-glass tube as above
2   syringe clamps, with bosses and stands, or syringe bench

Optional: a small balloon

Chemicals

Wire-form copper (II) oxide, laboratory reagent quality
Source of hydrogen

Preparation of wire-form copper powder: see experiment 1.1

Procedure

The apparatus is set up as in Fig. 1.1 with the silica tube filled with wire-form copper powder. There are two ways of filling one of the syringes with exhaled air. One is simply to blow through the three-way stopcock. The other is to blow up a small balloon and to attach it to the three-way stopcock by means of a rubber band. In both cases use the last breath of air to be expelled from the lungs to obtain the maximum carbon dioxide and minimum oxygen content.

From this point the procedure is exactly similar to that for experiment 1.1.

Procedure using simplified apparatus:

The procedure is similar to that in experiment 1.1 but the plastic syringe must be filled by exhaling air directly into it. The gas is prevented from escaping by holding a finger over the end and coupling it as soon as possible with the rest of the apparatus.

Results

Results vary considerably depending on the time the air has been in the lungs and other factors.

1.3 Determining the percentage of carbon dioxide in exhaled air

Introduction

This experiment follows from experiment 1.2. The two experiments may be combined, using the apparatus for experiment 1.2 together with the Orsat pipette used in this experiment. The experiment is suitable for pupil use or as a demonstration.

Apparatus

1   100 cm³ glass syringe (lightly greased)
1   glass tube bent through 90°
1   Orsat pipette
1   syringe clamp, with boss and stand, or syringe bench

Optional: balloon and three-way stopcock

Chemicals

5 M sodium hydroxide or potassium hydroxide (about 100 cm³)

Procedure

Set up the apparatus as shown in Fig. 1.3. Fill the syringe to 100 cm³ with exhaled air, either by direct blowing or by blowing up the balloon and attaching it to the three-way stop-cock with a rubber band and then filling the syringe through the stopcock.

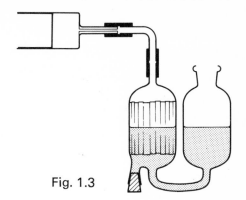

Fig. 1.3

Half fill the Orsat pipette with 5 M sodium hydroxide or potassium hydroxide. Pump the exhaled air from the syringe in and out of the Orsat pipette, taking care not to let any of the gas bubble through the pipette and escape to the atmosphere. Use the level of the solution in the Orsat pipette as a manometer and read the volume of gas in the syringe (see p. 4, use of Orsat pipette). Repeat the pumping action. When the volume of gas in the syringe remains constant for two readings you can take it that the carbon dioxide is all absorbed. The reduction of volume is equal to the volume of carbon dioxide absorbed.

Results

Results vary considerably depending on the time the air has been in the lungs and other factors.

1.4 The quantitative decomposition of mercury (II) oxide

Introduction

This experiment may be taken as an extension to Priestley's experiment (the thermal decomposition of mercury oxide). If it is established that the gas given off when 'red calx of mercury' is heated is oxygen what will we find if we measure the volume given off? The treatment will depend upon the stage in the chemistry course at which this experiment is performed. At an early stage it may be used to find the approximate density of the gas. The weight is given by the loss in weight of the mercury oxide; the volume by the increase in volume shown by the syringe. At a later stage it may be used to find the formula of mercury oxide (given atomic weights). The experiment is a simple one and is suitable for pupils to do themselves, and may also be performed with trilead tetroxide (red lead) or potassium permanganate.

Apparatus

1   100 cm³ glass syringe, or 1 50 cm³ plastic syringe
1   bent tube and bung to fit test-tube
1   hard glass test-tube
1   syringe clamp, with boss and stand

Chemical

Mercury (II) oxide

Fig. 1.4

Procedure

Weigh out about 0·5 g of mercury (II) oxide accurately (three decimal places) in the hard glass test-tube; weigh the test-tube empty and with the mercury oxide. Assemble the apparatus so that the test-tube is free to be moved by virtue of the flexibility of the rubber connecting tube. Heat the mercury oxide until no more oxygen is given off and the syringe ceases to move out. By this time all the mercury should be in the form of small beads round the middle or upper part of the tube. Allow the apparatus to cool to room temperature before recording the volume of the gas in the syringe. Note atmospheric pressure and temperature.

Results

Accurate results can be achieved for this experiment provided that the mercury oxide itself is pure. The volume of oxygen evolved should be correct within ±5%.

1.5   The decomposition of potassium chlorate and potassium perchlorate

Introduction

The results of these experiments may be used as an example of the Law of Multiple Proportions. Alternatively, both experiments may be used as steps in the determination of the formulae of the respective compounds.

Apparatus

1   100 cm³ glass syringe, or 1 50 cm³ plastic syringe
1   bent tube and bung to fit test-tube
1   hard glass test-tube
1   syringe clamp, with boss and stand

Chemicals

Analytical quality potassium chlorate and potassium perchlorate

Fig. 1.5

Procedure

The procedure used is very similar to that in experiment 1.4. Set up the apparatus as shown in Fig. 1.5 Weigh out about 0·2 g of the chlorate or perchlorate in the test-tube. Check that the apparatus is gas-tight and heat the test-tube gently at first and then strongly. When no more oxygen is evolved allow the apparatus to cool to room temperature and record the volume of oxygen produced. Repeat the experiment with the second compound.

Results

Results should be accurate to within a few per cent if the chemicals are pure and in accordance with the equation

$$2KClO_{3(s)} \rightarrow 2KCl_{(s)} + 3O_{2(g)}$$
$$KClO_2 \rightarrow KCl + 2O_2$$

1.6   The decomposition of sodium hydrogen carbonate

Introduction

This can be linked with experiment 1.8 to determine the formula of sodium hydrogen carbonate. The compound is first heated in a test-tube and the products – sodium carbonate, water and carbon dioxide – are analysed qualitatively. The quantitative experiment described below may then be attempted. This experiment is suitable for performance by pupils, or as a demonstration.

Apparatus

1. 100 cm³ glass syringe
1   bent tube (45°) and bung to fit
1   test-tube
1   syringe clamp, with boss and stand

Chemicals

Sodium hydrogen carbonate
Silica gel
Glass-wool

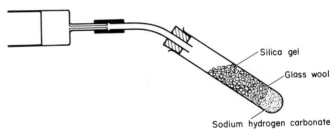

Silica gel

Glass wool

Sodium hydrogen carbonate

Fig. 1.6

Procedure

Weigh out about 2 millimoles of sodium hydrogen carbonate in the test-tube. Half fill the rest of the tube with a weighed amount of silica gel, separating it from the carbonate with a wad of glass-wool. Fix the test-tube to the syringe by means of the bung and bent tube. Heat the sodium hydrogen carbonate strongly but avoid heating the silica gel as far as possible. When decomposition is complete allow the apparatus to cool to room temperature. Record the volume of carbon dioxide evolved in the syringe. Pour out the silica gel and reweigh it to find the amount of water produced. The water will have been absorbed by the silica gel provided that it has not been heated too strongly. The residue may also be weighed. From this information the number of millimoles of water and carbon dioxide per millimole of sodium hydrogen carbonate may be calculated.

Results

Results tend to be a little low due to the absorption of carbon dioxide in the water. The expected volume of carbon dioxide is given by the equation $2\,NaHCO_{3(s)} \rightarrow Na_2CO_{3(s)} + CO_{2(g)} + H_2O_{(l)}$.

1.7   The action of acid on sodium carbonate and sodium hydrogen carbonate

Introduction

This simple experiment may be treated in several different ways. It may be used as part of the determination of the equation for the reaction of carbonates with hydrochloric acid (see Nuffield, O-level Chemistry Course, *The Basic Course*, topic 17). It may also be used as part of an investigation into the properties and formulae of carbonates and hydrogen carbonates (bicarbonates). Finally, it may be used to show that a mole of any carbonate or hydrogen carbonate gives the same volume of carbon dioxide when treated with an acid. The experiment may be performed with almost any carbonate – provided that the chloride is soluble.

Apparatus

1   100 cm³ glass syringe, or 1 50 cm³ plastic syringe
1   bung and bent tube (90°) to fit
1   boiling tube
1   syringe clamp, with boss and stand
1   ignition tube

Chemicals

5 M hydrochloric acid (about 15 cm³)
Anhydrous sodium carbonate
Anhydrous sodium hydrogen carbonate

Procedure

Pour about 15 cm³ of the 5 M hydrochloric acid into the boiling tube.

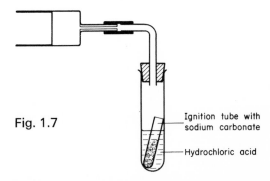

Fig. 1.7

Ignition tube with
sodium carbonate

Hydrochloric acid

Add a small quantity of the carbonate and shake the tube vigorously. This serves to saturate the solution with carbon dioxide. Weight about 2 millimoles of the carbonate or hydrogen carbonate in the ignition tube. Set up the apparatus as shown in Fig. 1.7 and drop the ignition tube containing the carbonate into the boiling tube, immediately closing the tube with the bung. Shake the boiling tube so that the acid splashes into the ignition tube and reacts with the carbonate. Record the volume of carbon dioxide in the syringe when evolution of the gas has ceased. Allow to cool to room temperature if necessary. Record room temperature and pressure.

Results

Results tend to be low if the acid is not thoroughly saturated with carbon dioxide. For this reason the best results are often obtained by repeating the experiment two or three times with the same acid.

1.8 The action of hydrochloric acid on magnesium and other metals

Introduction

This experiment is very similar to 1.7 in technique. It can be used to determine part of the equation for the reaction between magnesium and hydrochloric acid – how many moles of hydrogen are produced by one mole of magnesium? Teachers interested in the historical approach may like to use it as an experiment to find the equivalent weight of magnesium. The experiment may be used for any other metal which reacts with hydrochloric acid in a similar way. It is suitable for pupils' use.

Apparatus

1   100 cm³ glass syringe, or 1 50 cm³ plastic syringe
1   bung and bent tube (90°) to fit
1   boiling tube
1   syringe clamp, with boss and stand

Chemicals

Dilute hydrochloric acid
Concentrated hydrochloric acid
Magnesium ribbon

Procedure

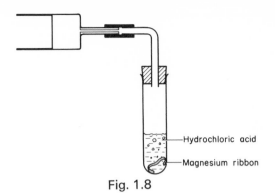

Fig. 1.8

Set up the apparatus shown in Fig. 1.8. Weigh out about 4 millimoles (0·048 g) of magnesium accurately. Put about 10 cm³ each of dilute and concentrated hydrochloric acid into the boiling tube. Tilt the boiling tube at an angle of about 45° and put the magnesium ribbon on the side of the tube taking care that it does not slide into the solution. Place the bung, which is connected to the syringe, into the boiling tube checking that there is no air leak before turning the boiling tube into the vertical position and thus allowing the magnesium ribbon to fall into the acid. After the reaction is over wait for the boiling tube and its contents to cool to room temperature before recording the volume of hydrogen evolved from the syringe.

Results

Results tend to be low if there is even the smallest leak in the apparatus. The best results are obtained when the rate of evolution of hydrogen is high.

1.9   The action of chlorine on metals

Introduction

This experiment can be used for determining the formulae of the chlorides – how many moles of chlorine react with one mole of the metal to produce the chloride? – or in the traditional way to find the equivalent weights of metals. Care must be taken with the use of chlorine and the experiment should only be performed by pupils if fume cupboards are available. Good results can be obtained with mercury, tin, zinc, iron, bismuth, antimony and lead. I have found it difficult to obtain reliable results from either aluminium or sodium.

Apparatus

2   100 cm³ glass syringes
1   silica-glass tube with a 2 cm diameter bulb blown in it (see
      diagram)
1   three-way stopcock
Apparatus for producing *dry* chlorine
2 syringe clamps, with bosses and stands, or syringe bench

Chemicals

Sample of the metal or metals to be used, in powder, granular or
   wire-form
Potassium permanganate and concentrated hydrochloric acid for
   producing chlorine
Glass-wool

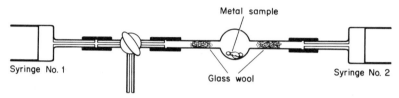

Fig. 1.9

Procedure

Weigh about 1 millimole of the metal to three or four decimal places
and put it in the bulb of the silica tube. Put a tuft of glass-wool
each side of the bulb. Set up the apparatus as shown in Fig. 1.9.
Flush out and fill syringe No. 1 with 100 cm³ of dry chlorine. The
chlorine may be prepared in the usual way from concentrated hydrochloric acid and potassium permanganate, but it *must* be carefully
dried. The gas may be dried by bubbling it through concentrated
sulphuric acid.

Pass about 20 cm³ of the chlorine from syringe No. 1 to syringe
No. 2 before heating the metal in the silica tube. This will ensure that
no air is in contact with the metal. This precaution should not be
taken with the tin as it sometimes leads to spontaneous ignition.
Now heat the metal carefully, passing the chlorine steadily to and fro
over the metal as heating continues. The glass-wool should prevent
most of the chloride from getting into the syringes. Allow the apparatus to cool at room temperature before measuring the volume of
chlorine which has reacted with the metal. Note room temperature
and pressure.

Results

Results for this experiment are good provided that the metal samples
are pure and the apparatus dry. The volume of chlorine which reacts
is within 5% of the expected volume. The higher oxidation state
chloride is produced in each case.

Chapter Two

Formulae and Equations

2.1 Finding the formula of ammonia

Introduction

This is an extremely neat experiment of the type which can only be done quickly and efficiently by the use of syringes. Using the three-syringe method it is suitable as a demonstration. It may be linked with a general study of ammonia and its place in fertilisers and food production. See the Nuffield O-level Chemistry Course, *The Basic Course*, topic 22. It also serves as an introduction to equilibria – see Chapter 4, especially experiment 4.7. The experiment can be simplified for pupils by reducing the number of syringes to two and, if necessary, using plastic syringes.

3   100 cm³ glass syringes
2   three-way stopcocks
2   silica tubes, with
4   2 cm glass rods to fit loosely in the tube
3   syringe clamps, with bosses and stands, or syringe bench

Apparatus for producing dry ammonia:

1   boiling tube
1   bung to fit boiling tube with hole to fit
1   drying tube
1   bung to fit drying tube with hole to fit
1   T-piece

Simplified apparatus for pupil use:

2   100 cm³ glass syringes
1   three-way stopcock
1   silica tube with 2 2 cm glass rods to fit loosely in the tube
2   syringe clamps, with bosses and stands

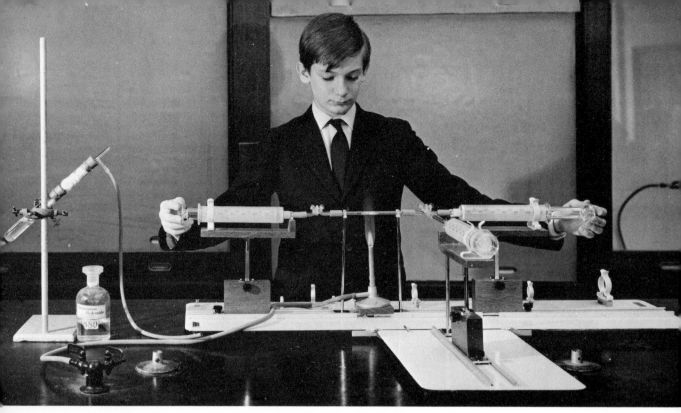

Fig. 2.1A

Chemicals

0·880 ammonia
Quicklime lumps for drying ammonia
Wire-form copper (II) oxide, laboratory reagent quality
Degreased iron wire-wool.

Procedure

Set up the apparatus as shown in Fig. 2.1B. Silica tube 'A' should be

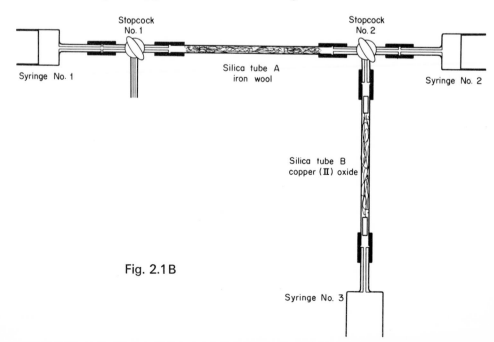

Fig. 2.1B

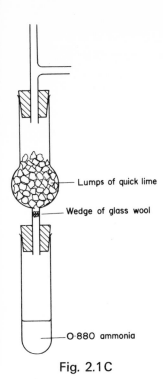

Lumps of quick lime

Wedge of glass wool

0·880 ammonia

Fig. 2.1C

filled with copper (II) oxide, plugged at each end with a glass rod, and silica tube 'B' should be filled with freshly reduced wire-wool, similarly plugged with two glass rods. It is essential for this experiment that the apparatus should be dry. For the best results, wash out the apparatus with nitrogen before starting the experiment. Fill syringe No. 1 through stopcock No. 1 with 40 cm³ of dry ammonia, having first washed it out two or three times with the gas. Turn the stopcocks so that syringe No. 1 is connected to syringe No. 2. Heat the wire-wool with a roaring Bunsen flame so that it becomes red-hot and pass the ammonia backwards and forwards across it. The gas will rapidly expand in volume as the ammonia decomposes into nitrogen and hydrogen. When there is no further increase in volume allow the gases to cool and measure the total volume. It should be about 80 cm³. Now push all the gas into syringe No. 2, turn stopcock No. 2 so that syringes Nos. 2 and 3 are connected, and heat the copper oxide with a Bunsen burner. Push the mixture of nitrogen and hydrogen backwards and forwards across the copper oxide until there is no further contraction in volume. The hydrogen will reduce the copper oxide to copper and water will be produced. Record the volume of the remaining nitrogen.

Possible sources of error

The ammonia must be dry. If either the ammonia or apparatus is damp the result of the volume of dissociated gases may be too high, as some ammonia will have dissolved in the water. If the flame is not hot enough full association will not occur and the value for the dissociated gases will be low.

Results

Results

Given the conditions described above 40 cm³ of ammonia should give between 79·5 and 80 cm³ of nitrogen and hydrogen. The final volume of nitrogen is always high and may be as much as 23 cm³. Some of this must be due to the water vapour present at this stage of the experiment. The equation for the reaction is

$$2NH_{3(g)} \rightarrow N_{2(g)} + 3H_{2(g)}.$$

2.2 The reaction between ammonia and hydrogen chloride

Introduction

This is a simple experiment which can easily be performed by pupils using, if desired, cheap plastic syringes. Better results can be obtained with glass syringes, but there is a tendency for the nozzles to become filled with solid ammonium chloride. The problem put to the pupils is: 'In what proportion by volume do ammonia and hydrogen chloride react?'

Apparatus

2   100 cm³ glass syringes, or 2 50 cm³ plastic syringes
1   three-way stopcock
2   syringe clamps, with bosses and stands, or syringe bench

Chemicals

Source of dry ammonia (see experiment 2.1)
Source of dry hydrogen chloride

Dry ammonia may be simply produced by heating 0·880 ammonia gently in a boiling tube with a drying tube containing calcium oxide attached – see diagram for experiment 2.1. Dry HCl may be produced by the traditional method of dropping concentrated sulphuric acid into a mixture of rock salt and sodium chloride crystals, or by dropping concentrated hydrochloric acid into concentrated sulphuric acid.

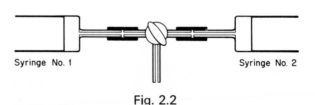

Syringe No. 1   Syringe No. 2

Fig. 2.2

Procedure

Set up apparatus as shown in Figure 2.2. Flush out and fill syringe No. 1 with 40 cm³ of dry ammonia through the three-way stopcock. Flush out and fill syringe No. 2 with 60 cm³ of dry hydrogen chloride in the same way. Pupils may bring their apparatus to the fume cupboard where the ammonia and hydrogen chloride can be dispensed under the supervision of the teacher. Having filled the syringes, turn the stopcock so that both syringes are isolated from each other and from the atmosphere.

Turn the stopcock to connect the two syringes with each other and inject 10 cm³ of hydrogen chloride at a time into the ammonia. Note the change in total volume each time. The total volume ceases to decrease when 40 cm³ have been added.

Results

Results are accurate for this experiment *provided that the gases are pure.* Make quite sure that no air is mixed with the gases. A one to one ratio of volumes is obtained

$$NH_{3(g)} + HCl_{(g)} \rightarrow NH_2Cl_{(s)}.$$

2.3 The reaction between nitrogen monoxide (nitric oxide) and oxygen

Introduction

This experiment may be done as a demonstration or by the pupils. It is capable of being treated in several different ways. At its simplest it may be used to find the relative volumes of nitrogen monoxide and oxygen which react, i.e. to find x and y in the equation:

$$x\mathrm{NO} + y\mathrm{O}_2 \rightarrow \ldots$$

A good way in which to do this is to inject nitrogen monoxide into oxygen 5 cm³ at a time and plot total volume against volume of oxygen added. The volume will cease to contract when the reaction is complete. Note that this is a relatively slow reaction and that the reaction mixture takes some time to equilibrate.

The product of the reaction is, of course, a mixture of nitrogen dioxide and dinitrogen tetroxide in equilibrium. This fact may be taken as an opportunity to discuss the small volume of the product, leading to the idea of association. The equilibrium may be studied by diluting the mixture with air or by putting the syringe in a furnace and studying the effect of temperature upon it. See experiment 4.5.

Apparatus

2   100 cm³ glass syringes
1   three-way stopcock
2   syringe clamps, with bosses and stands, or syringe bench
Apparatus for producing pure nitrogen monoxide
  (Copper and 50% nitric acid produce impure nitrogen monoxide. The Thiele method in which a solution of sodium nitrite is added to a mixture of iron (II) sulphate and concentrated hydrochloric acid gives relatively pure nitrogen monoxide. See *Lecture Experiments in Chemistry* by G. Fowles, p. 314, for details of this preparation. Alternatively, a cylinder of the gas may be used.)

Chemicals

Nitrogen monoxide (see above)

Oxygen. Cylinder or standard preparation
(It is possible also to use air in this experiment instead of pure oxygen.)

Procedure

Set up the apparatus as shown in Figure 2.3. Turn the three-way stopcock so that syringe No. 2 is connected to the atmosphere. Flush out syringe No. 2 two or three times with oxygen and fill it with 30 cm³ of the gas. Turn the stopcock to connect syringe No. 1 to the atmosphere and similarly flush out and fill the syringe with 80 cm³ of nitrogen monoxide. The total volume of nitrogen monoxide and oxygen together is now 110 cm³. Turn the three-way stopcock so that the two syringes are connected and inject 10 cm³ of the nitrogen monoxide into the oxygen, quickly closing the stopcock again.

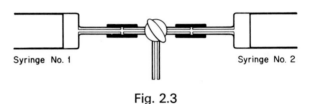

Syringe No. 1   Syringe No. 2

Fig. 2.3

Note the volume of the gases in syringe No. 2 and record the total volume of the gas in both syringes. Repeat this procedure until all the nitrogen monoxide has been injected into syringe No. 2. Now plot a graph of total volume against volume of nitrogen monoxide added. It will be found that the total volume decreases steadily until 60 cm³ of nitrogen monoxide is added. It then remains nearly constant. This shows that the reaction is complete when 60 cm³ of nitrogen monoxide has been added to 30 cm³ of oxygen. After this point a small increase in volume may be noticed. This is due to the dilution effect upon the equilibrium.

Results

The $2NO_2 \rightleftharpoons N_2O_4$ total volume at the 'end point' is about 32 cm³ showing that the NO_2 is largely associated (see experiment 4.5).

The reaction is surprisingly slow for a gaseous reaction. It is wise to wait at least three minutes between mixing the gases and reading their volume.

If the nitrogen monoxide is pure the 'end point' will not vary significantly from expected result of $2NO + O_2$. A typical set of results is given on p. 27.

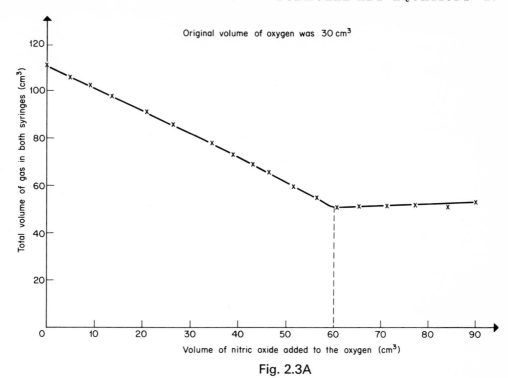

Fig. 2.3A

2.4 Dalton's experiment

Introduction

John Dalton claimed to have based his atomic theory upon the evidence (among other things) of an experiment with oxygen and nitrogen monoxide. Unfortunately he did not describe the experiment in sufficient detail to allow an exact repetition of the experiment to be performed. As a result of his experiment he concluded that nitrogen monoxide and oxygen could combine in two different proportions by volume, but in no other way. He performed two experiments. In one the nitrogen monoxide and oxygen were mixed in a tube and then the resulting gases absorbed by water. This experiment corresponds to experiment 2.3. If required, the volume of nitrogen dioxide and dinitrogen tetroxide produced in that experiment can be found by absorbing the gases in an Orsat pipette after every injection of nitrogen monoxide. Dalton's second experiment involved the rapid mixing of nitrogen monoxide and oxygen over a large surface of water. This is more difficult to simulate, but the following experiment, devised by W. J. Porterfield, provides very similar conditions.

Apparatus

1   1-litre flask
1   bung for the flask with 2 glass tubes through it (see diagram)
1   100 cm³ glass syringe
1   50 cm³ plastic syringe (or a second 100 cm³ glass syringe)
2   Mohr clips
1   bowl or pneumatic trough
2   syringe clamps, with bosses and stands, or syringe bench

Chemicals

Nitrogen monoxide (see experiment 2.3)
Oxygen (cylinder or standard preparation)

Procedure

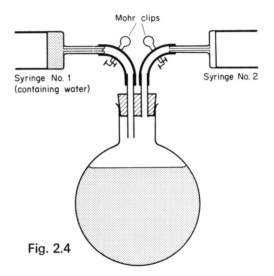

Fig. 2.4

Fill the flask with water and invert it (with the neck under water) in a bowl of water. Flush out and fill a syringe with 40 cm³ of nitrogen monoxide and bubble it into the flask. This may be done by attaching a rubber tube to the syringe, but if you use this method, make quite sure that there is no residual air in the tube. It is best to have the tube full of water in the bowl before it is attached to the syringe. Now flush out and fill the syringe with 10 cm³ of oxygen and bubble this into the flask, shaking the flask vigorously as soon as this is done. A brown gas is formed which quickly dissolves in the water. Repeat the addition of oxygen 10 cm³ at a time and in each case note the approximate change of volume. It will be found to decrease to almost nothing when 40 cm³ of oxygen are added and then increase if oxygen is in excess.

If air is used instead of oxygen or if greater accuracy is required the following method can be used to measure the residual volume of gas in the flask. Put the bung with the two tubes into the flask while it is under water and close the Mohr clips. Remove the flask from the bowl and turn it the right way up (see Fig. 2.4). Fill the plastic syringe (No. 1) full of water and attach it to one rubber tube. Attach the other syringe (No. 2), which must be closed, to the other rubber tube. Note that the tube to which this is attached only just reaches the bottom of the bung. Open the two Mohr clips and push the water from the plastic syringe into the flask. This will expel the gas into the glass syringe where its volume can be measured. A similar result is, of course, obtained with air. The reacting volumes being again about one part of nitrogen monoxide to one part of oxygen (five parts of air).

Results

I have never seen a really satisfactory explanation of the mechanism of this reaction and would welcome an explanation of it. Its interest is as much historical as chemical.

2.5 Finding the formula of dinitrogen monoxide (nitrous oxide)

Introduction

This experiment together with experiment 2.6 on nitrogen monoxide may be included in any topic on the formulae of compounds or in an investigation into the nature and properties of the oxides of nitrogen. It is simple to perform and gives reliable results. It is equally suitable as a pupil experiment or as a demonstration.

Apparatus

2 100 cm³ glass syringes, or 2 50 cm³ plastic syringes
1 silica-glass tube with 2 glass rods about 2 cm long to fit loosely in the tube.
1 three-way stopcock
2 syringe clamps, with bosses and stands, or syringe bench

Chemicals

Source of pure dinitrogen monoxide (see note below)
Wire-form copper powder (see experiment 1.1)

Preparation of pure dinitrogen monoxide:

This may be done conveniently by dropping a solution of hydroxylamine hydrochloride on to a hot, nearly saturated solution of iron (III) ammonium sulphate. Use solution containing about 10 g of hydroxylamine hydrochloride in 50 cm³ of water and 100 g of iron (III) ammonium sulphate in 200 cm³ of hot water.

Procedure

Fill the silica tube with 'wire-form' copper powder (see experiment 1.1, p. 10) and put the glass rods in either end to prevent the copper coming out into the syringes. Flush out and fill syringe No. 1 with 60 cm³ of dinitrogen monoxide. Turn the stopcock to connect syringe No. 1 with syringe No. 2 via the silica tube. Heat the copper in the

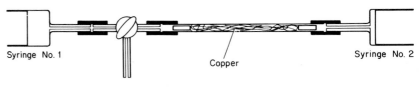

Syringe No. 1   Copper   Syringe No. 2

Fig. 2.5

silica tube with a moderate Bunsen burner flame and pass the gas backwards and forwards through the tube. The dinitrogen monoxide will decompose into nitrogen and oxygen, the oxygen reacting with the pink copper to form black copper (II) oxide. Cool the silica tube with a damp cloth before reading the volume of the remaining nitrogen. The volume will remain unaltered.

The formula of dinitrogen monoxide may be investigated more fully by weighing the silica tube and its contents before and after the experiment, and hence finding the weight of oxygen in the initial volume of dinitrogen monoxide.

Results

Results in this experiment should be accurate to within 5% or better. It is of course vital that the dinitrogen monoxide should be pure.

2.6   Finding the formula of nitrogen monoxide (nitric oxide)

Introduction

This experiment is similar to experiment 2.5. Both may be demonstrated at the same time or, if pupils are performing the experiment, half may be given dinitrogen monoxide and half nitrogen monoxide to analyse. They should not find it difficult to suggest copper as a reducing agent for the experiment. After the experiment the discussion of results should prove stimulating.

Apparatus

2   100 cm³ glass syringes
1   silica-glass tube with 2 glass rods about 2 cm long to fit loosely in the tube
1   three-way stopcock
2   syringe clamps, with bosses and stands, or syringe bench

Chemicals

Source of pure nitrogen monoxide (see experiment 2.3)
Wire-form copper powder (see experiment 1.1)

Procedure

Syringe No. 1   Syringe No. 2

Copper

Fig. 2.6

The procedure is exactly the same as that for experiment 2.5. In this case the volume of nitrogen left will be exactly half the volume of the nitrogen monoxide initially in the syringe.

Once again the silica tube containing copper may be weighed before and after the experiment. If the same volume of dinitrogen monoxide and nitrogen monoxide respectively was used in the two experiments, the increase in weight of the tube will also be the same, showing that each gas contains the same number of atoms of oxygen per molecule.

Results

Similar accuracy to that in experiment 2.5 is to be expected provided that the nitrogen monoxide is pure.

2.7   The reduction of carbon dioxide by carbon

Introduction

This experiment may be performed to determine the equation for the reaction between carbon and carbon dioxide, or may be used as evidence for Gay-Lussac's Law and hence Avogadro's hypothesis. The main difficulty with the experiment lies in achieving a sufficiently high temperature. Two roaring Bunsen burners may be just sufficient. Electrical heating by means of nickel chromium wire is, however, more effective (see p. 6). It is important to use the right quality of charcoal. Ordinary wood charcoal does not give good results, as it tends to evolve gases at about 1,000 °C. Granulated bone charcoal is better. The 'gas free' type should be specified.

Apparatus

2   100 cm³ glass syringes
1   three-way stopcock
1   silica-glass tube, optionally wound with nichrome wire for heating electrically
Glass-wool
2   syringe clamps, with bosses and stands, or syringe bench

Chemicals

Pure dry carbon dioxide, from a cylinder or by the standard method of acid on carbonate
Dry granulated bone charcoal (gas free)
Nitrogen, preferably from a cylinder (optional)

Procedure

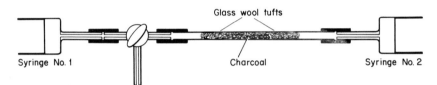

Fig. 2.7

Half fill the silica-glass tube with charcoal, keeping it in position in the centre of the tube with tufts of glass-wool. Make sure that the whole apparatus is dry. For the best results flush out the whole apparatus with nitrogen first. Then flush out and fill syringe No. 2 with 40 cm³ of carbon dioxide through the three-way stopcock. Turn the stopcock to connect the two syringes via the silica tube.

Heat the silica tube very strongly with two roaring Bunsen burners or with the electrical heating system. When the charcoal is white-hot pass the carbon dioxide slowly backwards and forwards through the silica tube until it ceases to expand. On cooling it should be found to have doubled in volume.

Test the carbon dioxide by expelling through a tube attached to the three-way stopcock and showing that it burns with a luminous blue flame.

The gaseous part of the equation $CO_{(g)} + C_{(s)} \rightleftharpoons 2CO_{(g)}$ may then be deduced. The charcoal and silica tube may be weighed before and after the experiment to determine the quantity of carbon in the equation.

The experiment could be extended by oxidising the carbon monoxide to carbon dioxide again using copper (II) oxide.

Results

Success in this experiment depends on getting the charcoal white-hot. The reaction is in equilibrium and results will tend to be a little low.

2.8   Finding the formula of hydrogen chloride

Introduction

This experiment may be used as a straight determination of formula or as an example of Gay-Lussac's Law of Combining Volumes. As the mixture of hydrogen and chlorine is explosive it is suitable as a demonstration experiment only. Some teachers have reported accidents with this experiment. The author has performed the experiment on numerous occasions without ever having an explosion. It is probable that explosions have been caused by insufficient packing of the glass-wool in the silica tube (see below). The glass-wool should be well packed, but must allow the passage of gases. Note that *10 cm* of the silica-glass tube should be packed with glass-wool.

Apparatus

2   100 cm³ glass syringes
3   three-way stopcocks
1   manometer, filled with light oil
1   Orsat pipette
1   silica-glass tube, 15 cm by 7 to 8 mm diameter
1   platinum spiral made from about 6 cm of 0·2 mm diameter
     wire. It should just fit in the silica-glass tube
Glass-wool
2   syringe clamps, with bosses and stands, or syringe bench
Safety screen

Chemicals

Dry hydrogen, preferably from a cylinder, alternatively by standard laboratory preparation.
Dry chlorine. Cylinders are seldom available in schools. The standard method of preparation by the action of concentrated hydrochloric acid on potassium permanganate may be used. To purify the gas, bubble it first through water and then through concentrated sulphuric acid.

Procedure

Pack the silica-glass tube with glass-wool for 10 cm of its length. The glass-wool should be fairly tightly packed but gases must be able to pass freely through it. The platinum spiral is placed in the open end of the tube and a glass rod put in after it to keep it in position (see Fig. 2.8).

Set up the apparatus as shown in Fig. 2.8, with the Orsat pipette

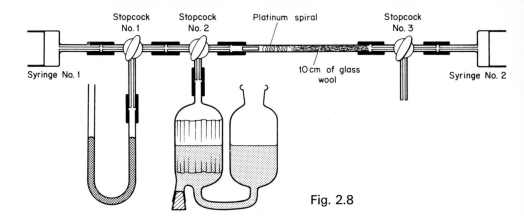

Fig. 2.8

half full of water. *The apparatus must be completely dry.* Flush out and fill syringe No. 1 with 50 cm³ of hydrogen from the cylinder via three-way stopcock No. 3. The flushing-out process is particularly important in this experiment and should be carried out three times. Turn the stopcock to connect syringe No. 2 to the atmosphere and flush out and fill it with 20 cm³ of chlorine. Once again it is advisable to flush the syringe out three times. (It is sensible to do this part of the experiment in a fume cupboard.) Place a black cloth over syringe No. 2 and turn stopcock No. 3 so that syringe No. 1 and syringe No. 2 are connected. Push the hydrogen from syringe No. 1 into the chlorine in syringe No. 2. The mixture is potentially explosive in the presence of sunlight, but the black cloth will prevent a reaction from starting. Place a safety screen between the apparatus and the class. A plastic safety screen is made which fits over the silica tube. Keep pupils away from the line of the syringes.

Heat the silica-glass tube until the platinum spiral is red-hot and then pass the mixture of hydrogen and chlorine very slowly from syringe No. 2 over the spiral to syringe No. 1. Most of the hydrogen and chlorine will combine during the first pass but to ensure that the reaction is complete, the tube should be cooled, the gases passed back into syringe No. 2 and the procedure repeated once or twice. By this time all the chlorine will have been converted into hydrogen chloride. Note the volume of the gases after the reaction. To find the volume of hydrogen chloride turn stopcock No. 2 so that syringe No. 2, containing the gases, is connected with the Orsat pipette. Pump the gases in and out of the Orsat pipette until no further decrease in volume takes place. Flush the gases backwards and forwards between the two syringes to collect any residual hydrogen chloride and repeat the process of absorption. Use the manometer to check that the gas is at atmospheric pressure before noting its volume. The hydrogen chloride is very readily absorbed by water and the decrease in volume denotes the volume of hydrogen chloride in the mixture. The remaining gas is of course hydrogen. (This may be confirmed if necessary.) All the chlorine has reacted (this may also be confirmed), the volume of hydrogen which reacted may be found by subtracting the remaining volume from the original volume of hydrogen (60 cm³) and the volume of hydrogen chloride is the decrease in volume on absorption. Approximately equal volumes of hydrogen and chlorine are found to react to produce double the volume of hydrogen chloride. Assuming the diatomicity of hydrogen and chlorine we have

$$H_{2(g)} + Cl_{2(g)} \longrightarrow 2HCl_{(g)}$$

Results

Some hydrogen chloride is always absorbed by the moisture in the apparatus however carefully it is dried. This results in a slight contraction of volume after the reaction. Thus 20 cm³ of chlorine and 60 cm³ of hydrogen produce about 79 cm³ of hydrogen chloride. This contracts to about 40 cm³ ($\pm$ 1 cm³) when absorbed over water.

Errors will be caused if the apparatus is not dry. As hydrogen is being used it is particularly important to see that the apparatus is gas-tight.

2.9 The volume composition of water

Introduction

This is best performed as a demonstration experiment. It is a quick and simple way of determining the volume composition of water. It can, of course, also be used as an illustration of Gay-Lussac's Law of Combining Volumes. The best results are obtained by repeating the experiment several times. Enough water then collects in the combustion pipette to allow it to be drained by means of the tap at the bottom of the pipette and tested.

Apparatus

2   100 cm³ glass syringes
1   combustion pipette, with 2-volt battery
2   three-way stopcocks
1   protective screen
2   syringe clamps, with bosses and stands, or syringe bench
Manometer filled with light oil (optional)

Chemicals

Hydrogen, from a cylinder, or if this is not available by the standard preparation using zinc and hydrochloric acid. The hydrogen need not be dry, but it must not contain any oxygen.

Procedure

Set up the apparatus as shown in Fig. 2.9 putting the transparent protective screen over the combustion pipette. Make sure that the coil in the combustion pipette is directly opposite the inlet tube and not more than 2 mm away from it. Flush out and fill syringe No. 2 with 40 cm³ of hydrogen via stopcock No. 2. It is absolutely vital that there is no residue of air in the syringe. The syringe should therefore be flushed out carefully at least three times. If the hydrogen is being generated chemically the usual test should be made to ensure that it is free of oxygen before filling the syringe. Turn stopcock No. 2 so that all three connections are 'off', i.e. the combustion pipette and the syringe are both cut off from the atmosphere. Now draw 40 cm³ of air into syringe No. 1 through the stopcock on the combustion pipette. Make sure that stopcock No. 1 is set so that the syringe and the combustion pipette are connected, but the manometer is cut off.

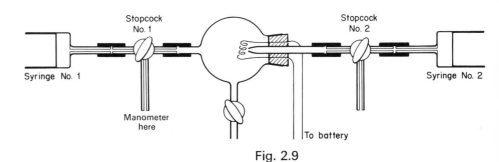

Fig. 2.9

Connect the battery and wait until the coil of nickel chromium wire is glowing red-hot. Then turn stopcock No. 2 to connect syringe No. 2 with the combustion pipette and *very gently and slowly* push the hydrogen into the combustion pipette. The hydrogen will ignite; the flame should be about 1 centimetre long. When all the hydrogen has burnt disconnect the battery and allow the apparatus to cool down. A damp cloth will help to cool the combustion pipette. When the apparatus is cool turn stopcock No. 1 so that the manometer is connected with the rest of the apparatus and adjust syringe No. 1 until the manometer shows that the pressure inside the apparatus is atmospheric. Read the volume of syringe No. 1. It should show a decrease of 20 cm³ demonstrating the left-hand side of the equation:

$$2H_{2(g)} + O_{2(g)} \rightarrow 2H_2O_{(l)}$$

Results

Better results are often obtained on repeating the experiment two or three times; it is quick to perform once the apparatus is set up. The errors may be due to the effect of water vapour on the volume of the gases.

Fig. 2.10A

2.10   Determining the formula of carbon monoxide

Introduction

Like the previous experiment this one makes use of the combustion pipette. The teaching approach may be 'How can we find the formula of this oxide of carbon?' or it may be to find the equation for the burning of carbon monoxide. For the first, the formula of carbon dioxide and the atomicity of oxygen must be assumed. For the second, the formula of carbon monoxide must also be assumed. It may be argued that if the formulae are known the equation can be deduced – this is not completely true and in any case experimental verification is always necessary. This experiment has the particular advantage that it is entirely gaseous. It is best done as a demonstration.

Apparatus

2   100 cm³ glass syringes
3   three-way stopcocks
1   Orsat pipette or diffusion absorber (see p. 4)
1   combustion pipette, with 2-volt battery
1   protective screen

2   syringe clamps, with bosses and stands, or syringe bench
Manometer filled with light oil

Chemicals

Carbon monoxide. This is best produced by the action of concentrated sulphuric acid on formic acid
Approximately 5 M potassium hydroxide solution for the Orsat
pipette

Procedure

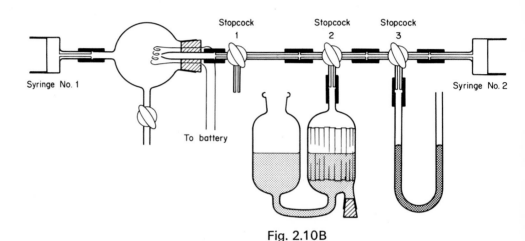

Fig. 2.10B

Set up the apparatus as shown in Fig. 2.10B. Syringe No. 2 should be
greased with a little Vaseline or similar lubricant. The Orsat pipette
is half-filled with a concentrated solution of potassium hydroxide
Place the protective screen over the combustion pipette.

Open all three stopcocks to ensure that the whole apparatus is at
atmospheric pressure. Then turn stopcocks Nos. 2 and 3 so that the
manometer and Orsat pipette respectively are cut off, but syringe
No. 1 is connected to the atmosphere via stopcock No. 1. Draw
20 cm³ of air into syringe No. 1 and turn stopcock No. 1 so that it is
closed to the atmosphere.

Flush out and fill syringe No. 2 with 40 cm³ of carbon monoxide via
stopcock No. 1. As with the hydrogen experiment the syringe must
be well flushed out and there must be no air mixed with the carbon
monoxide. Turn stopcock No. 1 so that all three tubes are cut off.
Connect the battery to the heating spiral in the combustion pipette
and wait until the spiral is red-hot. Then turn stopcock No. 1 to
connect syringe No. 2 with the combustion pipette and *gently and
slowly* push the carbon monoxide into the pipette. The gas should
burn with a flame about 1 cm long. When all the carbon monoxide
has been burnt allow the apparatus to cool. A damp cloth placed on
the combustion pipette will speed up the process. When the apparatus
is cool measure the increase in volume shown by syringe No. 2. This

should be about 20 cm³, showing a decrease in the *total* volume of gases of half the volume of carbon monoxide.

To estimate the volume of carbon dioxide present, use is made of the Orsat pipette. Push the gases from syringe No. 1 to syringe No. 2. Turn stopcock No. 2 so that syringe No. 2 and the Orsat pipette are connected. Pump the gases down into the Orsat pipette for a short time to allow the carbon dioxide to be absorbed. Take care not to allow any gas to bubble through the Orsat pipette and escape. Repeat this until no more gas is absorbed, noting the volume by which the gases have decreased. The decrease in volume will equal the volume of carbon dioxide absorbed. There is a danger that gas may leak through the syringe at this stage as a slight excess pressure is involved. This is why syringe No. 2 is greased. As the syringe is greased it is preferable to use the manometer to check that the volume of the gases is measured at atmospheric pressure. The manometer may be joined by means of a short length of rubber tubing to stopcock No. 2. The stopcock should only be opened to the manometer at the time of measuring the pressure of the gas in the syringe. If a manometer is not used it is possible to judge the pressure with reasonable accuracy from the level of the potassium hydroxide solution either side of the Orsat pipette.

There may be carbon dioxide remaining in the combustion pipette. To deal with this draw in 50 cm³ of air through the tap in the combustion pipette to syringe No. 2. This will bring some of the carbon dioxide from the combustion pipette. The same procedure as before is used to determine the volume of carbon dioxide present. After this a second volume of air may be drawn in the same way. The procedure should be repeated until there is more carbon dioxide left in the combustion pipette.

The volume of carbon dioxide evolved and the decrease of volume of the gases on burning will enable the formula of the gas or the equation of the reaction to be determined:

$$2CO_{(g)} + O_{2(g)} \rightarrow 2CO_{2(g)}$$

Results

The expected contraction of volume should be obtained without difficulty in this experiment provided that the apparatus is cooled to room temperature. The carbon dioxide volume obtained by absorption tends to be a little low; presumbably due to incomplete absorption. Some patience is needed.

2.11   The formulae of methane and other gaseous hydrocarbons

Introduction

Use of the combustion pipette allows the determination of the formula of any gaseous hydrocarbon. The method is essentially the same as that for experiment 2.10, the formula of carbon monoxide. The gas is burnt in air and the decrease in total volume is noted. The volume of carbon dioxide produced is then found by absorption.

Let the formula of the hydrocarbon be $C_x H_y$, then its reaction with oxygen is:

$$C_x H_y + \left(x + \frac{y}{4}\right) O_2 \rightarrow xCO_2 + \frac{y}{2}H_2O$$

That is to say 1 volume of the hydrocarbon reacts with $\left(x + \frac{y}{4}\right)$

volumes of oxygen to produce x volumes of carbon dioxide. The water is a liquid under the conditions of the experiment and its volume is negligible. When 1 volume of the hydrocarbon is burnt therefore the contraction in *total* volume is:

$$1 + x + \frac{y}{4} - x$$

$$= 1 + \frac{y}{4} \text{ volumes}$$

The absorption of carbon dioxide by potassium hydroxide solution will give x. Thus both unknowns are found.

It is worth going through this method of calculation before the experiment is done. It is possible to buy cylinders of hydrocarbons and both butane and propane are available in small cylinders for camping and for cigarette lighter fuels. The cylinders may be covered or painted over and the experiment treated as an investigation. This experiment may be linked with others such as the determination of molecular weight by the effusion method, experiment 5.5, page 72.

Apparatus

2   100 cm³ glass syringes
3   three-way stopcocks
1   Orsat pipette or diffusion absorber (see p. 4)
1   combustion pipette, with 2-volt battery
1   protective screen
2   syringe clamps, with bosses and stands, or syringe bench
Manometer filled with light oil

Chemicals

Methane or other hydrocarbon gas
Approximately 5 M potassium hydroxide solution for the Orsat
   pipette

Procedure

The procedure is essentially the same as that for experiment 2.10 except that methane or some other gaseous hydrocarbon is used instead of carbon monoxide. The same precautions must be taken. Results are accurate enough to determine the whole numbers in the formula.

For the lower molecular weight hydrocarbons (methane to propane) 40 cm³ of the gas should be burnt from syringe No. 1. Make sure that syringe No. 2 contains enough air to allow for the final contraction. 50 cm³ of air will be sufficient for the lighter gases.

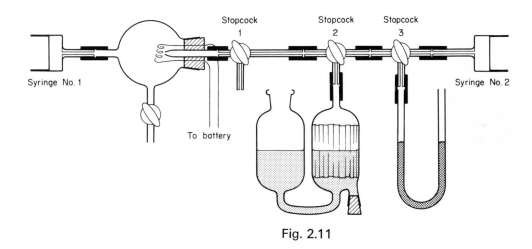

Fig. 2.11

Heavier gases present problems as the contraction in volume is greater and allowance must also be made for the fact that there is an *increase* in volume while the hydrocarbon is burning. This can be managed by having an additional syringe attached by a three-way stopcock next to syringe No. 2, and also by cooling the combustion pipette with a damp cloth.

Results

Results are sufficiently accurate in all cases for the determination of the formulae.

2.12 The chlorination of unsaturated hydrocarbons

Introduction

This experiment is suitable for any gaseous compound containing a carbon-carbon double bond. The addition reaction is simple and quantitative. A similar reaction may be performed with acetylene and its derivatives, but there is a distinct danger of explosion in this case and the experiment should only be performed by the very experienced or the foolhardy.

The experiment with ethylene and its derivatives is safe if carried out as described below. It is particularly useful as part of the structural determination of ethylene or as a step in the investigation into the structure of an 'unknown' gas. Butene or propene are possible examples.

The experiment is suitable either as a demonstration or for class use.

Apparatus

2   100 cm³ glass syringes
1   three-way stopcock
2   syringe clamps, with bosses and stands, or syringe bench

Chemicals

Ethylene (or other unsaturated hydrocarbon gas)
(Ethylene may be prepared on a small scale by the action of excess concentrated sulphuric acid on ethanol)
Dry chlorine. Chlorine may be prepared by the conventional method of the action of concentrated hydrochloric acid on potassium permanganate.

Procedure

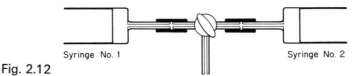

Syringe No. 1   Syringe No. 2

Fig. 2.12

Set up the apparatus in a fume cupboard where the chlorine and ethylene can be produced without danger. Flush out and fill syringe No. 1 with 50 cm³ of ethylene. It is important to ensure that no air is left in the syringe; the flushing out should therefore be performed two or three times. Turn the stopcock to connect syringe No. 2 with the atmosphere and similarly flush it out two or three times before filling it with 20 cm³ of chlorine. The apparatus may now be removed from the fume cupboard.

Turn the stopcock to connect the two syringes and push the chlorine into the ethylene. Oily drops of dichloroethane will be seen to form immediately and the volume will decrease by 20 cm³. This shows that the reaction between ethylene and chlorine is equimolar.

$$C_2H_{4(g)} + Cl_{2(g)} \rightarrow C_2H_4Cl_{2(g)}$$

In the case of other gaseous hydrocarbons the decrease in volume will show how many double bonds the hydrocarbon contains.

Results

The results for this experiment are accurate to within the accuracy of the syringe, provided that the gases are pure.

2.13   The formula of hydrogen sulphide

Introduction

This experiment is based on the reaction between heated copper and hydrogen sulphide:

$$Cu_{(s)} + H_2S_{(g)} \rightarrow CuS_{(s)} + H_{2(g)}$$

The hydrogen sulphide must be pure, but in every other respect the experiment is very simple. The experiment is suitable for class use or as a demonstration. As with many of the other experiments in this chapter, it may be used simply to find the formula of hydrogen sulphide, as an example of Gay-Lussac's Law of Combining Volumes, or as a reaction whose equation is to be determined.

As a weighing is involved in this experiment, the use of the 'millimole' is advantageous. A millimole (one-thousandth of a mole) of any gas at N.T.P. occupies about 22·4 millilitres (cm³); at room temperature and pressure about 24 cm³. A millimole of sulphur weighs 32 milligrammes. In the following experiment we find that 1 millimole of hydrogen sulphide gives, on decomposition with copper, 1 millimole of hydrogen (H) and 1 millimole of sulphur (combined with the copper). Assuming the diatomicity of hydrogen the formula of hydrogen sulphide as H_2S follows from these results.

Apparatus

2   100 cm³ syringes
1   three-way stopcock
1   silica-glass tube with 2 glass rods of about 2 cm length and a diameter just smaller than the inside diameter of the tube
2   syringe clamps, with bosses and stands, or syringe bench

Chemicals
Copper, made by reducing (wire-form) copper oxide. (See experiment 1.1)

Pure dry hydrogen sulphide
(This is best made by the action of dilute hydrochloric acid on antimony (III) sulphide. Iron (II) sulphide gives an impure product. Calcium or barium sulphide may also be used.)

Procedure

Fig. 2.13

Fill the silica tube with 'wire-form' copper and put the glass rods in either end to stop the copper from coming out. Weigh the tube and its contents. Set up the apparatus as shown in Fig. 2.13. Flush out and fill syringe No. 1 with 48 cm³ (2 millimoles) of hydrogen sulphide. Turn the stopcock so that the two syringes are connected through the silica tube. Heat the copper in the silica tube with a Bunsen burner until the reaction is complete (i.e. no more copper sulphide is formed). Remove the Bunsen burner and cool the silica tube with a damp cloth. When the tube is cool, measure the volume of hydrogen remaining. The volume should be 48 cm³ (2 millimoles). The gas may be tested by pushing it out into an inverted test-tube and 'popping' it in a Bunsen burner flame. Remove the silica tube and its contents and weigh it. The gain in weight is equal to the weight of sulphur in the hydrogen sulphide. It should be about 64 mg (2 millimoles).

Results
The only problem with this experiment concerns the purity of the hydrogen sulphide. If the gas is pure the change in volume after cooling should be less than 2%.

2.14 The reaction between chlorine and hydrogen sulphide

This reaction is very similar to that of experiment 2.7. It may be used as the basis for an investigation. 'How does chlorine react with hydrogen sulphide?' A qualitative investigation will show that hydrogen chloride and sulphur are the products of the reaction. 'What is the equation for the reaction?' The answer can be guessed from the nature of the products but can only be confirmed by experiment.

'How shall we do the experiment?' Pupils should find it easy to suggest a simple method like the one described below. It is unwise to allow the hydrogen sulphide and chlorine to react inside one of the syringes as the sulphur deposited jams the syringe and is difficult to remove. The experiment is suitable as a demonstration or for pupil use.

Apparatus

2 100 cm³ glass syringes
2 three-way stopcocks
1 100 cm³ (or 150 cm³) conical flask, with rubber bung and inlet tubes as shown in the diagram.
2 syringe clamps, with bosses and stands, or syringe bench

Chemicals

Pure dry hydrogen sulphide (see p. 44)
Pure dry chlorine

Procedure

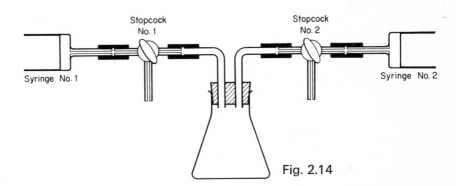

Fig. 2.14

Set up the apparatus as shown in Fig. 2.14. Ensure that the apparatus is completely dry. Fill the flask and syringe No. 1 with pure dry hydrogen sulphide through stopcock No. 2. This should be done in a fume cupboard. To ensure that all the air is driven out of the flask, stopcock No. 1 can be opened and hydrogen sulphide allowed to flow through the flask for a time. When this has been done syringe No. 1 should be flushed out and filled with 40 cm³ of hydrogen sulphide. Now flush out and fill syringe No. 2 with 30 cm³ of chlorine. Turn both the stopcocks so that the syringes are connected via the flask and push the chlorine into the flask. After allowing a few minutes for the gases to mix thoroughly, measure the volume shown by syringe No. 1. It should now show 70 cm³. That is to say there should be no change in total volume.

Results

In fact there is often a small decrease in total volume as some of the hydrogen chloride is absorbed by the small amount of water adsorbed to the glass. Better results are obtained if the experiment is repeated two or three times in the same apparatus.

The experiment establishes the left-hand side of the equation:

$$H_2S_{(g)} + Cl_{2(g)} \rightarrow 2HCl_{(g)} + S_{(s)}$$

2.15   The oxidation and reduction of copper (I) oxide

Introduction

This simple experiment may be used to determine the formulae of copper (I) and copper (II) oxides. It acts as an example of the law of multiple proportions. The only difficulty concerns the purity of copper (I) oxide. Commercially available material tends to contain an excess of oxygen as the compound reacts with oxygen if it is left in contact with the air. It can be 'purified' by heating to red heat in a stream of carbon dioxide or nitrogen.

Apparatus

2   100 cm³ glass syringes
1   silica-glass tube
1   three-way stopcock
2   syringe clamps, with bosses and stands, or syringe bench

Chemicals

Copper (I) oxide, analytical quantity
Cylinder of hydrogen, or some other source of hydrogen
Glass-wool
Cylinder of nitrogen or carbon dioxide for 'purifying' the copper (I) oxide

Procedure

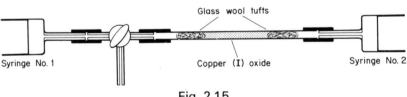

Fig. 2.15

Weigh the silica-glass tube with two tufts of glass-wool in it. Put between 150 and 200 mg of pure copper (I) oxide in the silica tube, keeping the powder in position by means of the tufts of glass-wool, and weigh it again. Set up the apparatus as shown in the diagram. Check that the apparatus is gas-tight. Draw in 100 cm³ of air into syringe No. 1. Heat the copper (I) oxide gently and pass the air backwards and forwards over it until no further decrease in volume occurs. Allow the apparatus to cool, the silica tube may be cooled with a damp cloth, and note the decrease in volume.

Now flush out the apparatus several times with hydrogen finally leaving 100 cm³ in syringe No. 1. Heat the copper oxide again, this time passing the hydrogen over it until all the oxide has been reduced to copper. Cool the apparatus and note the decrease in volume of the hydrogen.

Results

The decrease in volume of the hydrogen will be about four times the decrease in volume of the oxygen. This will make a good discussion point with the class.

red copper oxide + 1 vol. of oxygen ⟶ black copper oxide
black copper oxide — 4 vols. of hydrogen ⟶ copper

If necessary the composition by weight of the two oxides can be determined, the density of oxygen being assumed.

The final volume of hydrogen is liable to be too high as it is saturated with water vapour. This error can be eliminated by saturating the oxygen and the hydrogen with water first. For this purpose put a drop or two of water in the silica tube and heat the tube while passing the gases through it.

2.16 The formula of sulphur trioxide

Introduction

This experiment may be demonstrated or performed as a class experiment. It is a good example of the use of a catalyst as well as being a reaction of industrial importance. It may also be used as another example of Gay-Lussac's law.

Apparatus

2   100 cm³ glass syringes
1   silica-glass tube with glass rods to fit inside
1   three-way stopcock
2   syringe clamps, with bosses and stands, or syringe bench

Chemicals

Source of oxygen (preferably a cylinder)
Source of sulphur dioxide (preferably a cylinder)
Platinised asbestos

Procedure

Syringe No. 1   Platinized asbestos   Syringe No. 2

Fig. 2.16

Set up the apparatus as in the diagram. In this case do *not* dry the apparatus or the platinised asbestos too carefully. There must be some water to absorb the sulphur trioxide. On the other hand do not deliberately wet the apparatus or the asbestos as this will cause other complications.

Flush out and fill one syringe with 40 cm³ of sulphur dioxide and the other with 60 cm³ of oxygen. Mix the gases and pass them over the platinised asbestos which must be heated to red heat.

White fumes appear but are soon absorbed by the moisture on the surface of the glass in the cool parts of the apparatus. After a time allow the apparatus to cool and note the reduction of volume. It will be about 20 cm³. The residual gas will be found to be oxygen (glowing splint) although traces of sulphur dioxide are also present. Wash the apparatus out carefully after the experiment.

Results

Results should be consistent with the equation

$$2SO_{2(g)} + O_{2(g)} \rightarrow 2SO_{3(s)}$$
within 5%.

Chapter Three

Organic Analysis

3.1   Finding the formulae of liquid hydrocarbons

Introduction

The experiments in this chapter are well worth performing as they represent the only simple methods available in schools for determining the formulae of organic compounds. The techniques are nevertheless more difficult than those of the preceding chapters and it is strongly recommended that the techniques are practised carefully before being demonstrated in front of a class. The experiments are suitable for better pupils who have time to spend on them; say in the science club or in voluntary laboratory time. I have used them for class-work in the first-year sixth.

The formulae of liquid hydrocarbons are relatively easy to work out and several approaches may be made depending on the intelligence of the pupils and the objectives of the experiment. At its simplest it is possible to assume that the 'unknown' compound is a saturated hydrocarbon. It is, of course, much more satisfactory to let the pupils do preliminary experiments with the liquid to show that this is the case. Now one can ask them how they can find out which hydrocarbon it is. The simple answer is to burn it and to determine the volume of carbon dioxide produced from a known weight of the liquid. Assuming the formula C_nH_{2n+2} for saturated hydrocarbon the hydrogen content can easily be calculated once the carbon content is known. Alternatively, a slightly more sophisticated experiment may be performed in which the hydrogen content is measured by collecting and weighing the water produced when the hydrocarbon is oxidised.

The experiment is also suitable for use with liquid unsaturated hydrocarbons. The hydrogen content must then be measured.

Apparatus

2   100 cm³ glass syringes
1   silica-glass tube, with 1 2 cm glass rod to fit inside
2   three-way stopcocks

1   U-tube with side arms (optional)
1   1 cm³ syringe with needle
2   syringe clamps, with bosses and stands, or syringe bench

Chemicals

Liquid hydrocarbon
Asbestos wool (or other inert absorbent material)
Wire-form copper (II) oxide
Silica gel for the U-tube
Cylinder of nitrogen, if available

Procedure

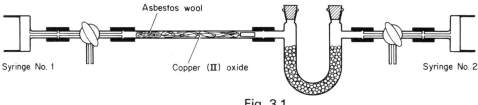

Fig. 3.1

Set up the apparatus as shown in Fig. 3.1. The silica-glass tube should contain wire-form copper (II) oxide packed for three-quarters of its length, a glass rod being inserted in the right-hand end to prevent copper oxide from getting into the drying tube or syringe, and in the left-hand end to keep the asbestos wool in place. Insert a small plug of loosely packed asbestos wool into the other end of the silica tube. If a cylinder of nitrogen is available, flush out the apparatus with nitrogen through the three-way stopcock, leaving about 40 cm³ in syringe No. 1. If no nitrogen is available put some copper (made by reducing wire-form copper (II) oxide) in the silica tube in place of some of the copper oxide, draw in 50 cm³ of air and pass it over the heated copper until all the oxygen has been removed from the air. It is not essential to remove the oxygen from the air, but the calculation is clearer to the pupils if this is done.

Use the 1 cm³ syringe and needle to inject a known quantity of the hydrocarbon on to the asbestos wool. The actual quantity will vary with the hydrocarbon used, but about a third of a millimole is right for hexane. The volume of carbon dioxide produced will then be 2 millimoles; about 50 cm³. If some 40 cm³ of nitrogen (or air) is left in the left-hand syringe the total final volume of gases will be about 90 cm³.

If the weight of water produced is going to be measured fill the U-tube with silica gel and weigh it. Make quite sure that the bungs in the top of the U-tube are gas-tight. Test the whole apparatus to see that it is gas-tight. Now fill the 1 cm³ syringe with about the correct volume of the liquid hydrocarbon, weigh it on a three-place balance, inject the liquid on to the asbestos wool, and reweigh the syringe. The weight of hydrocarbon injected on to the asbestos wool is thus known.

Heat the copper oxide strongly with a Bunsen burner until it is red-hot. To ensure that hydrocarbon vapour does not move towards syringe No. 1 draw nitrogen through the silica tube *very slowly* by pulling out syringe No. 2. Start this as soon as the copper wire begins to get hot. The heat from the red-hot copper oxide will probably be enough to vaporise the hydrocarbon. If it is not, warm the asbestos wool gently until the hydrocarbon vaporises. Continue to draw the nitrogen (or air) very slowly from syringe No. 1 to syringe No. 2. The gas carries the vapour through the red-hot copper oxide where it is oxidised to carbon dioxide and water. To make sure that oxidation is complete, pass the carrier gas backwards and forwards several times through the red-hot copper oxide. The first pass is of course the most important one; if hydrocarbon vapour comes through the copper oxide unchanged it is likely to condense in the rest of the apparatus. Allow the apparatus to cool to room temperature and record the increase of volume of the gases in the syringes. If nitrogen has been used the increase in volume is equal to the volume of carbon dioxide produced. If air has been used allowance must be made for the fact that one-fifth of the volume of the air (the oxygen) has been changed into carbon dioxide. It is of course possible to fit an Orsat pipette filled with concentrated potassium hydroxide and to find the volume of carbon dioxide by absorption.

If the U-tube has been used this must be weighed after the experiment.

Better results are obtained for the carbon dioxide if the U-tube is not used. Two experiments may be performed. One without the U-tube to determine the carbon dioxide and hence the carbon content and one with the U-tube to determine the water and hence the hydrogen content.

Results

The treatment of results can be varied considerably (see Introduction).

The simplest possible method is to determine the molecular weight first (see Chapter Five) and then use this to work out how many millimoles of the liquid have been oxidised. The increase in the volume of gases in the apparatus will show exactly how many millimoles of carbon dioxide have been produced. The number of millimoles of carbon dioxide produced per millimole of liquid is equal to the number of carbon atoms in the molecule. Thus hexane will give 6 millimoles of carbon dioxide per millimole of hexane.

The gain in weight of the U-tube is, of course, due to water absorbed. The weight of hydrogen in the hydrocarbon is equal to one-eighth of the increase in weight of the U-tube.

An accuracy within about 10% is to be expected from this experiment. The use of an electric heater, made by winding nichrome wire round the silica tube, improves the results. Better results are obtained with the more volatile liquids.

3.2 The action of sodium on alcohol

Introduction

This experiment may be used with experiment 3.3 as part of a determination of the full structure of an alcohol such as ethanol. The first stage is to give the pupils some of the unknown liquid and tell them to find out what they can about it. With some help they will find that it burns to form carbon dioxide and water and perhaps that it can be oxidised to other compounds (acetaldehyde, acetic acid). They can then be told to investigate its reaction with sodium (with due precautions). Having found that a white solid, presumably ionic, and hydrogen are produced, this experiment can be performed as one designed to find out how much hydrogen is produced from a given weight of alcohol. As in experiment 3.1 the calculation is greatly simplified if the molecular weight is already known. It is advisable, therefore, to do experiment 5.2 or 5.3 first. Once the molecular weight is determined the question becomes 'how many millimoles of hydrogen are displaced by sodium from 1 millimole of the alcohol?' Half a millimole of gaseous hydrogen will be replaced for every atom which is replaced in the molecule. The result of the experiment shows that one atom of hydrogen only is replaced. The experiment is suitable for pupils to perform.

Apparatus

1 100 cm³ glass syringe
1 boiling tube
1 bung (to fit boiling tube) with bent tube
1 syringe clamp, with boss and stand
1 1 cm³ glass or plastic syringe with needle
1 100 cm³ beaker

Chemicals

Alcohol
Sodium – freshly cut slices
Hexane or heptane, or if these are not available, 'paraffin'

Procedure

Prepare in advance a series of boiling tubes (one per pupil or pair of pupils) in which is placed about 4 cm depth of hexane (or heptane or paraffin) and about five slices of freshly cut sodium. Close the boiling tubes with bungs.

Set up the experiment as shown in Fig. 3.2. As a preliminary boil the liquid gently in the test-tube for about five minutes. This will drive off any dissolved gases and cause any water which may be present to react with the sodium. Cool the boiling tube and liquid to room temperature by placing it in a 100 cm³ beaker of water and discharge the gases produced from the syringe. A curious point about this experiment is that if the above process is repeated *some* gas is always given off. It is possible that the paraffins are 'cracked' to some small extent under these conditions.

Draw in an appropriate volume of alcohol into the 1 cm³ syringe and weigh the syringe and alcohol to three decimal places. Removing the bung from the boiling tube, quickly inject the alcohol on to the paraffin and replace the bung. Weigh the 1 cm³ syringe again and note the weight of alcohol injected. Make quite sure that the bung is firmly on the boiling tube and that the apparatus is gas-tight. The alcohol will begin to react with the sodium but the reaction is too slow unless

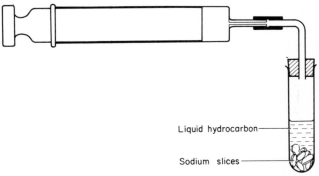

Fig. 3.2

the boiling tube is shaken quite vigorously and the paraffin heated to near its boiling-point. Gentle boiling may be allowed. After about ten minutes cool the contents of the boiling tube by placing them, as before, in a beaker of water. When the whole apparatus has reached room temperature, record the volume of gas evolved in the syringe. The process may be repeated to see if the reaction is complete.

Results

Under these conditions the reaction does not appear to go to completion. Results are usually some 10% low. The best results are obtained with the lower alcohols.

3.3   Finding the formulae of liquid alcohols

Introduction

This experiment may be used as part of a series (see the introduction to experiment 3.2). The molecular weight is established (experiments 5.2 and 5.3) and the fact that one atom of hydrogen per molecule of alcohol can be displaced by sodium. Now the molecular formula is found. The procedure used is similar to that of experiment 3.1, only the calculation is different.

Apparatus

2   100 cm³ glass syringes
1   silica-glass tube, with 1 2-cm glass rod to fit inside
2   three-way stopcocks
1   U-tube with side arms (optional)
1   1 cm³ syringe with needle
2   syringe clamps, with bosses and stands, or syringe bench

Chemicals

Alcohol
Asbestos wool (or other inert absorbent material)
Wire-form copper (II) oxide
Silica gel
Cylinder of nitrogen, if available

Procedure

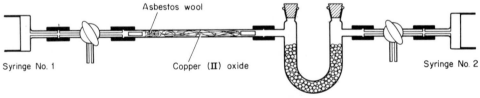

Fig. 3.3

The apparatus is set up as shown in Fig. 3.3 and the procedure given for experiment 3.1 is followed. Once again the total increase in volume gives the volume of carbon dioxide and hence the number of carbon atoms in the molecule. The gain in weight of the U-tube gives the weight of water produced and hence the number of hydrogen atoms in the molecule. The oxygen content is determined by subtraction (the assumption being made that no other element is present).

Results

The results for carbon and hydrogen content should be within about 10% of the known value, although a larger error is allowable. The lower alcohols give better results. As with experiment 3.1 some care and practice is needed to obtain good results. The use of electrical heating (winding a coil of nichrome wire round the silica tubes) produces a more even heat and gives the best results.

Chapter Four

Rates of Reaction and Equilibria

4.1 The rate of the reactions between carbonates and hydrochloric acid

Introduction

This experiment is suitable as an introduction to rates of reaction. The calcium carbonate reaction has the advantage that it is thermally almost neutral and therefore no noticeable change in temperature of the reactants occurs during the reaction. Questions which may be asked are: Does the size of the particles affect the rate of reaction? Does the concentration of hydrochloric acid affect the rate of reaction? Does the temperature affect the rate of reaction? The experiment may be treated as an investigation and different pupils or pairs of pupils be given the task of answering each question. The technique for the experiment is extremely simple. Plastic syringes do not run smoothly enough for use in this experiment. The cheaper type of glass syringe is quite satisfactory.

Apparatus

1   100 cm³ glass syringe
1   boiling tube
1   bung to fit boiling tube, with glass tube with 90° bend
1   syringe clamp, with boss and stand
1   porcelain boat (or ignition tube)
1   stop-clock or watch
1   thermometer (0–100 °C)

Chemicals

Carbonate or hydrogen carbonate, e.g. calcium carbonate, sodium carbonate (anhydrous), potassium carbonate, etc.
Hydrochloric acid of known strength (say 2 M)

Procedure

Set up the apparatus as shown in Fig. 4.1. Weigh out the required quantity of carbonate in a porcelain boat. Put about 20 cm³ of hydro-

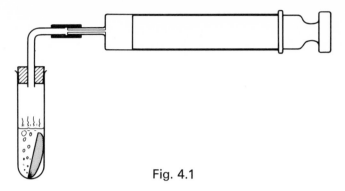

Fig. 4.1

chloric acid in the boiling tube. Clearly the volume used must remain constant for any series of experiments. The strength of the acid may be varied to find out how this affects the rate of reaction. The strength needed to produce a reasonable rate of reaction will vary according to the particle size of the carbonate and the particular carbonate used. The right strength must be found initially by trial and error. As carbon dioxide is moderately soluble in hydrochloric acid solution, it is *absolutely essential* to saturate the solution first. This is best done, before starting the experiment, by adding a spatulaful of the carbonate to the acid in the boiling tube and stirring it vigorously. When this has been done place the boat and carbonate on the side of the tilted boiling tube and immediately replace the bung in the boiling tube. Then tip the carbonate into the acid. At the same time start the stop-clock. Record the volume change in the syringe at regular intervals. After the experiment measure the temperature of the reactants again. Results are invalid if there has been a change of temperature of more than a degree or two. To keep the temperature constant immerse the boiling tube in a beaker of water.

Results

Good results can be obtained from this experiment if the suitable concentrations are found. It can be treated qualitatively only if necessary. The results are best displayed in graphical form.

4.2 The rate of decomposition of hydrogen peroxide

Introduction

This experiment is similar to experiment 4.1. It has the advantage that it allows the effect of a catalyst to be studied. Concentration and temperature can also be varied as before. In addition, such questions as: Is the catalyst used up during the reaction? Does the particle size of the catalyst affect the rate of reaction? and Which catalyst is most

effective? may be asked. A particularly interesting experiment may
be performed by performing the reaction once and then adding a
further dose of the manganese dioxide catalyst. Contrary to expecta-
tions, the reaction begins again. Presumably the catalyst is 'poisoned'
by the reaction. There are many possible investigations here. Once
again the apparatus is simple and the experiment suitable for pupils
to perform. Plastic syringes do not run smoothly enough for us in this
experiment. The cheaper type of glass syringe is quite satisfactory.

Apparatus

1 100 cm³ glass syringe
1 boiling tube
1 bung to fit boiling tube with glass tube bent through 90°
1 porcelain boat
1 syringe clamp, with boss and stand
1 stop-clock or watch
1 thermometer (0–100 °C)

Chemicals

Hydrogen peroxide of known strength
Manganese dioxide – granular and powder forms
Other metal oxides – particularly transition metal oxides such as
    copper oxide
Transitional metal powders – if available

Procedure

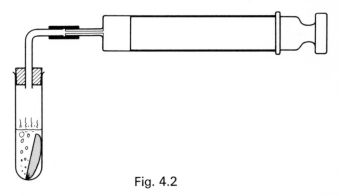

Fig. 4.2

The procedure is very similar to that of experiment 4.1. The apparatus
is set up as in Fig. 4.2 with 20 to 30 cm³ of hydrogen peroxide solution
in the boiling tube. The exact strength will have to be found, as
before, by trial and error. As a rough guide 10 volume hydrogen
peroxide decomposes at a reasonable rate when catalysed by granular
manganese dioxide. The powdered manganese dioxide causes a much
more rapid decomposition. Record the temperature of the hydrogen
peroxide. Weight out a gramme of the catalyst and drop it into the

hydrogen peroxide, quickly replacing the bung. Alternatively, tilt the boiling tube on its side, put the porcelain boat and catalyst carefully on the side without allowing it to slide into the hydrogen peroxide and replace the rubber bung. Now twist the boiling tube into the vertical position so that the boat and catalyst fall into the solution. At the same time start the stop-clock. Record the volume change at regular intervals. Check the temperature of the reaction.

The decomposition of hydrogen peroxide under these conditions is exothermic. Presumably this is due to the high bond energy of gaseous oxygen. This means that the reaction tends to heat up if the concentration of the hydrogen peroxide is at all high. To keep the temperature constant immerse the boiling tube in a beaker of water.

Perform a series of experiments varying one of the following conditions: temperature, concentration of hydrogen peroxide, type of catalyst, particle size of a particular catalyst.

Results

Good results are obtained for these reactions if the temperature is kept constant. The results are best expressed graphically.

4.3 The rate of reaction between a metal and hydrochloric acid

Introduction

These reactions afford another simple introduction to reaction rates. Once again the effect of temperature and concentration can be measured. In particular if magnesium ribbon or tin-foil is used an accurate measurement of the effect of the surface area of the reaction can be made.

Apparatus

1   100 cm³ glass syringe
1   boiling tube
1   bung to fit boiling tube with glass tube bent through 90°
1   syringe clamp, with boss and stand
1   thermometer
1   100 cm³ beaker

Chemicals

Metals such as magnesium (ribbon) and tin (foil)
Hydrochloric acid of known strength (1 or 2 M)

Procedure

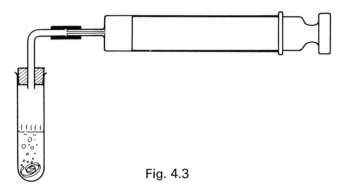

Fig. 4.3

Set up the apparatus with 20 to 30 cm³ of hydrochloric acid in the boiling tube. As in experiments 4.1 and 4.2 the most suitable strength will have to be found by trial and error. Place a measured length (or weighed piece) of magnesium ribbon on the side of the tilted boiling tube so that it is not in contact with the acid. Replace the bung and turn the boiling tube into the vertical position allowing the magnesium to slide into the acid. At the same time start the stop-clock. Record the volume of hydrogen evolved at regular intervals. Repeat the experiment for different lengths of magnesium ribbon. With tin-foil adopt a similar procedure but record the surface area for each experiment. As these reactions are exothermic, care must be taken to cool the contents of the boiling tube by placing it in a beaker of water.

With dilute solutions this is less important. Measure the temperature of the hydrochloric acid before and after the experiment to check that no significant increase in temperature has taken place.

Results

The results are best plotted graphically. After a time the magnesium ribbon or tin-foil will have reacted to such an extent that the surface area has altered considerably. The first part of the reaction is therefore the part to be used when determining the variation of reaction rates with surface areas.

4.4 Rate experiments with the gas syringe recorder

Introduction

The experiments described in the first chapter may be performed in an elegant way with the use of the gas syringe recorder. Minor modification in apparatus and technique are required and are described below. The apparatus is suitable for demonstration or for pupil use.

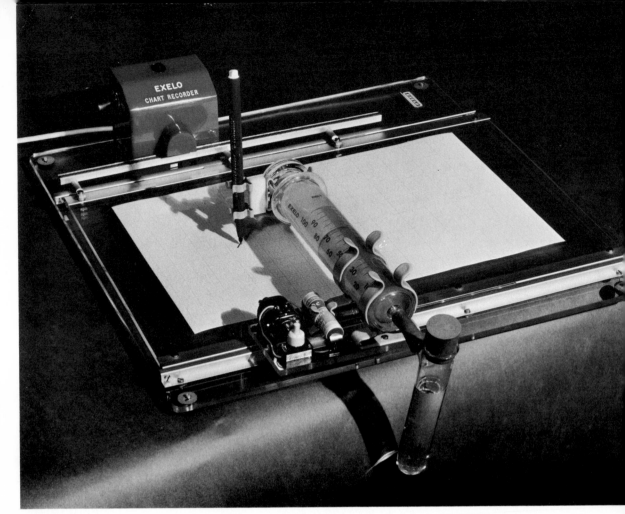

Fig. 4.4

The results are produced on paper or can be projected by an overhead projector.

Additional apparatus

Test-tube with side arm (or boiling tube with bung and bent tube, as before)
Bung to fit test-tube
Gas syringe recorder
The stop-clock will not be needed

Procedure

Set up the apparatus as shown in Fig. 4.4. Use essentially the same technique substituting the side arm test-tube for the boiling tube. Reduce the quantity of solution if necessary. Instructions for the use of the gas syringe recorder are supplied with the apparatus. The advantage of this apparatus is that a graph of volume of gas evolved against time is automatically produced.

4.5 The equilibrium between nitrogen dioxide and dinitrogen tetroxide

$$2NO_2 \rightleftharpoons N_2O_4$$

Introduction

This experiment follows on from experiment 2.3. The nitrogen dioxide, dinitrogen tetroxide equilibrium mixture being best prepared by the method described in that experiment. Note that for this experiment the syringe containing the mixture of gases should be lightly greased. There are two effects which may now be investigated. The first is the effect of dilution; the second the effect of temperature. The effect of dilution on the scale possible with syringes is relatively small but is interesting as a basis for discussion. The effect of temperature is much larger and leads to a relationship between the equilibrium constant and temperature which is much easier to obtain than by any other method. In particular the use of gases ensures that the equilibrium state is quickly reached.

The time taken for a reaction in liquid solution to equilibrate is a matter of hours or days, whereas in the gaseous state equilibrium is reached in a minute or two.

Apparatus

2   100 cm³ glass syringes (one greased)
1   three-way stopcock
1   syringe furnace or thermostatically controlled water bath
1   thermometer marked in 0·1 °C

Chemicals

Source of nitrogen dioxide. See experiment 2.3

Procedure

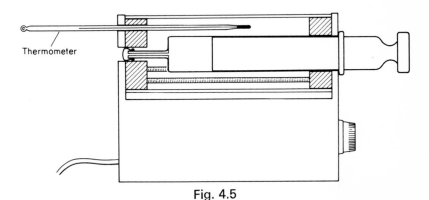

Thermometer

Fig. 4.5

Fill a greased syringe with 50 cm³ of the nitrogen dioxide/dinitrogen tetroxide mixture, as described in experiment 2.3. Immerse the syringe completely in a thermostatically-controlled water bath (a syringe clamp will·be needed to keep it under water) or place it in a syringe furnace of either of the two designs available (see p. 6). At high temperatures the syringe tends to leak so that the electric-bulb furnace is probably better. A thermostatically-controlled water bath gives a very even temperature distribution and gives the best results.

Record the volume of the gaseous equilibrium at a series of temperatures between room temperature and about 80 °C. Higher temperatures can be reached with the electric furnaces.

Results

The volume of unassociated NO_2 can be found from the volume of NO used to prepare the equilibrium mixture, using the equation:

$$2NO + O_2 \rightarrow 2NO_2$$

The equilibrium constant for the reaction:

$$N_2O_4 \rightleftharpoons 2NO_2$$

is given by

$$Kp = \frac{(pNO_2)^2}{pN_2O_4}$$

where pNO_2 and pN_2O_4 are the partial pressures of NO_2 and N_2O_4 respectively. Note that it is simpler to consider this equilibrium in terms of the dissociation of dinitrogen tetroxide.

If $\log Kp$ is plotted against $\frac{1}{T}$ ($°K$), the result will be a straight line showing that

$$\frac{1}{T} \propto \log Kp$$

which is consistent with the relationship

$$\Delta G° = - RT \log K.$$

The slope of the graph can thus be used to find a value for $\Delta G°$.

4.6   The dimerisation of acetic acid in the vapour phase

$$2CH_3COOH \rightleftharpoons (CH_3COOH)_2$$

Introduction

This experiment is an alternative to experiment 4.4 which can be used to study equilibria. It suffers from the disadvantage that it must be performed at higher temperatures. An electric syringe furnace must therefore be used. Results are not quite so precise as those with NO_2, but the experiment provides an interesting investigation. Just above its boiling-point (188 °C) acetic acid vapour is dimerised. At 300 °C only the monomer is present.

Apparatus

1   100 cm³ glass syringe (greased)
1   vaccine cap
1   syringe furnace and regulator
1   1 cm³ syringe and needle
1   0–300 °C thermometer

Chemicals

Acetic acid

Procedure

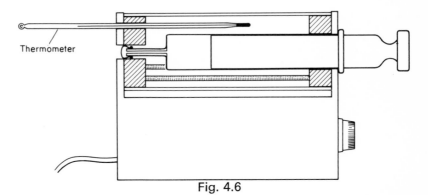

Thermometer

Fig. 4.6

Place the glass syringe in the furnace as shown in Fig. 4.6. Note that the nozzle of the syringe, with the vaccine cap fitted, must be well inside the furnace so that it is as hot as the rest of the syringe. Adjust the regulator so that the syringe furnace heats up to about 130 °C and *allow the syringe and furnace to equilibrate*. Draw up about 2 milli-moles of acetic acid into the 1 cm³ syringe, weigh it on a three-place balance and inject the acetic acid through the vaccine cap into the heated syringe. The acetic acid will vaporise immediately. Withdraw the 1 cm³ syringe and weigh it again. The difference in weight gives the weight of acetic acid injected. After allowing the vapour time to equilibrate (up to five minutes) note its volume. Raise the temperature of the furnace by 20 °C and, after giving time to equilibrate again note the volume of the vapour. Repeat this process up to 200 or 300 °C. Note atmospheric pressure.

Results

The treatment is similar to that in experiment 4.5. It is probably better simply to plot volume and then degree of dissociation against temperature as accuracy over this temperature range is not so good as in experiment 4.5.

4.7   The equilibrium between nitrogen, hydrogen and ammonia – The synthesis of ammonia

Introduction

The experiment to determine the formula of ammonia (experiment 2.1) can also be used as a qualitative illustration of equilibria. It will be found that when the ammonia is dissociated in experiment 2.1 the residual gases still contain a detectable quantity of ammonia. It is possible to reverse the process and produce a very small quantity of ammonia from nitrogen and hydrogen.

Apparatus

2   100 cm³ glass syringes
1   silica-glass tube
1   three-way stopcock
2   syringe clamps, with bosses and stands, or syringe bench

Chemicals

Hydrogen cylinder (or other source)
Nitrogen cylinder (or other source)
Freshly reduced iron wool (or I.C.I. catalyst 35–4)
Indicator paper

Procedure

A suitable beginning for this investigation is experiment 2.1. Having performed this experiment, show that the remaining gases contain ammonia. There is usually a sufficient ammonia residue to be detected by smell. Eject the gases through a stopcock on to a piece of damp red litmus (or universal indicator) paper.

Flush out the apparatus with nitrogen (or hydrogen) several times.

Each time heat the iron wool, passing nitrogen backwards and forwards two or three times. Then eject the gas over damp red litmus paper. Repeat the process until the litmus paper does not change colour. Now fill the apparatus with 60 cm³ of hydrogen and 20 cm³ of nitrogen. Pass the mixture of gases two or three times over the

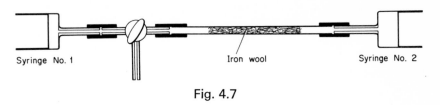

Syringe No. 1   Iron wool   Syringe No. 2

Fig. 4.7

strongly heated (red-hot) iron wool. Eject the mixture of gases, as before, over damp red litmus paper. Just enough ammonia (or other alkaline gas?) is evolved to turn a small part of the litmus paper blue. The percentage of ammonia theoretically produced under these conditions is less than $0 \cdot 1 \%$. There is no proof that the gas produced is in fact ammonia, but repeated experiments would seem to show that it is at least probable that it is!

Use of the I.C.I. catalyst to synthesise ammonia

The experiment is more convincing if the I.C.I. catalyst is used. The catalyst is based on magnetite, Fe_3O_4, which must first be reduced in a stream of hydrogen at 400–500 °C. This is best done in the silica-glass tube. The catalyst must not be exposed to the air after reduction. Fill the apparatus as before with 60 cm³ of hydrogen and 20 cm³ of nitrogen. Heat the catalyst to very dull red heat. It must not be heated too strongly. Pass the gases backwards and forwards over the catalyst several times. If the gases are then ejected over a piece of damp red litmus paper it will go blue. Once again this is an indication of the presence of ammonia rather than proof.

Chapter Five

The Molecular Weight of Gases and Volatile Liquids

5.1   The molecular weight of gases by direct weighing

Introduction

It is clearly desirable that the introduction to molecular weights in the fifth or sixth form should be made through experiments performed by the pupils themselves. It is also important that such experiments should be simple, direct and quick to perform. The experiments described in this chapter fulfil these criteria and provide a useful addition to experiments traditionally performed in schools.

The first experiment is designed as an introduction to the molecular weight of gases. It follows from Avogadro's hypothesis that the molecular weights of gases can be compared by direct weighing. Cannizzaro showed that the hydrogen molecule is diatomic. We can therefore compare the weight of a given volume gas with the weight of the same volume of hydrogen at the same temperature and pressure. Given that the molecular weight of hydrogen is 2:

$$\frac{\text{molecular weight of gas}}{2} = \frac{\text{weight of given volume of gas}}{\text{weight of given volume of } H_2}$$

The plastic syringe, which is light and gas-tight, is suitable for this experiment. It has the advantage that it can be weighed 'empty'; for details see the procedure. The technique described below was developed by Mr K. W. Badman.

Apparatus

1   plastic syringe, 20 or 50 cm³
1   piece of rubber tubing to fit nozzle (about 3 cm long)
1   Mohr clip
1   piece of wood to hold syringe open (see diagram)

Chemicals

Cylinders (or other sources) of the gases whose molecular weights are to be determined.

Procedure

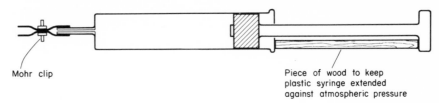

Fig. 5.1

Before weighing the gas, weigh the syringe 'empty'. In order that the buoyancy of the syringe should be the same it is necessary to put the rubber tube over the syringe nozzle and effectively close the syringe with a Mohr clip. Now pull the syringe open so that there is a vacuum of 20 cm³ (or 50 cm³) in it. If it is now released it will of course spring back to its original position. To keep it open place the piece of wood between the 'shoulder' of the barrel of the syringe and the top of the plunger (see Fig. 5.1). The syringe can now be weighed 'empty'. A four-place balance is preferable but not essential.

Fill the syringe with the gas whose molecular weight is to be determined. The syringe must be flushed out at least three times with the gas before finally filling it. Weigh the syringe and the gas.

Results

An accuracy to within 5% is possible for most gases. Hydrogen, being so light, gives the least satisfactory results.

5.2 The molecular weight of volatile liquids using a boiling liquid heater

Introduction

This experiment and the two following are modern versions of the Victor Meyer experiment. They rest on the theory of the direct weighing experiments but the actual weighing of the substance is performed, for convenience, in the liquid rather than the gaseous phase. The syringe provides an outstandingly simple method for doing this.

Apparatus

1   100 cm³ glass syringe, or 1 50 cm³ plastic syringe
1   vaccine cap to fit syringe
1   1 litre 'tall form' beaker
1   tripod and gauze
1   1 cm³ syringe and needle
1   thermometer

Chemicals

Liquid whose molecular weight is to be determined
Glycerol – or other high boiling-point liquid – for use with liquids
with boiling-points higher than 85 °C

Procedure

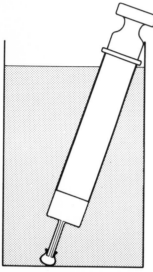

Fig. 5.2

For liquids with boiling-points less than 85 °C use water. For liquids
with boiling-points over 85 °C use a high boiling-point liquid such as
glycerol. Due care must be taken that the vapour does not catch fire,
the liquid must not be allowed to boil, and the experiment in this case
is best performed in a fume cupboard.

For pupil use the water experiments are simplest. Three-quarters
fill the 1 litre beaker with water and heat it to boiling on a tripod and
gauze. Put a vaccine cap on the syringe having drawn in 20 cm³ of
air. Place the syringe vertically in the beaker with the nozzle down-
wards so that all the graduations are covered with water. Allow the
apparatus a few minutes to equilibrate. Fill the 1 cm³ syringe and
needle with an appropriate quantity of the liquid whose molecular
weight is to be determined. About 2 millimoles will be sufficient.
Weigh the syringe with its contents to three decimal places. Remove
the 100 cm³ syringe from the boiling water (you will need to hold it
in a cloth or handkerchief) and inject the liquid into it through the
vaccine cap. Replace the 100 cm³ syringe in the boiling water and
reweigh the 1 cm³ syringe. The difference in weights gives the weight
of liquid injected. After a few minutes note the new volume of gases
in the syringe. The increase in volume of the gases gives the volume
of the vapour. Note the temperature of the boiling liquid and
atmospheric pressure.

Results

Results tend to be a little low. Results will certainly be low if the
vapour is too near its boiling-point.

5.3   The molecular weight of a volatile liquid using a furnace

Introduction

This experiment is very similar to experiment 5.2. A weighed quantity
of liquid is again vaporised, its volume being measured in a syringe.
The use of a furnace instead of a liquid heater makes possible a
greater range of temperatures, and hence the molecular weights of
a larger number of substances can be determined. The furnaces are
also cleaner to use than boiling liquids.

Fig. 5.3A

Apparatus	1	100 cm³ glass syringe
	1	vaccine cap
	1	syringe furnace
	1	thermometer

Apparatus

1   100 cm³ glass syringe
1   vaccine cap
1   syringe furnace
1   thermometer

Chemicals

Liquid whose molecular weight is to be determined

Procedure

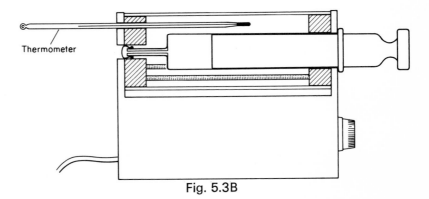

Thermometer

Fig. 5.3B

Place the syringe in the furnace with the vaccine cap over the nozzle. There should be no air in the syringe. Ensure that the nozzle is well within the furnace, so that it reaches the same temperature as the rest of the syringe. The temperature of the furnace should be at least 15 °C above the boiling-point of the liquid. Allow five minutes for the syringe to reach the same temperature as the furnace. Draw up about 2 millimoles of the liquid whose molecular weight is to be determined into the 1 cm³ syringe and weigh the syringe and liquid. Inject the contents through the vaccine cap into the heated 100 cm³ syringe.

After a few minutes read the volume indicated on the syringe. Note the temperature of the furnace and atmospheric pressure.

Results

Results tend to be a few per cent low. The technique described in experiment 5.4 may also be used.

5.4   The molecular weight of a volatile liquid using a light-bulb oven

Introduction

This method of determining molecular weights was developed by Mr K. W. Badman. The light-bulb oven may be bought from W. G. Flaig & Sons Ltd. or made in the laboratory. See the *School Science Review*, Vol. 49, No. 169, p. 904. The only limitation is that the maximum temperature of the furnace is about 120 °C. The method described here may also be used with the electric furnace (see experiment 5.3).

Apparatus

1   100 cm³ glass syringe
Vaccine cap to fit syringe
1   1 cm³ glass syringe with needle
1   light-bulb syringe oven

Procedure

Put a vaccine cap on the end of 100 cm³ glass syringe and place the syringe in the furnace. Switch the furnace on and allow at least five minutes for it to reach a steady temperature.

The essence of this technique is that a substance of known molecular weight is used as a standard. Acetone is suitable. Fill the 1 cm³ glass syringe with acetone, weigh it, and inject just enough acetone through the vaccine cap into the 100 cm³ syringe in the furnace to

Fig. 5.4A

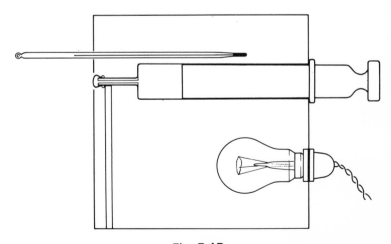

Fig. 5.4B

produce 60 cm³ of vapour. If too much acetone is injected it is even possible to withdraw it back into the 1 cm³ syringe where it will re-condense. Weigh the small syringe again to see what weight of acetone has been injected. Remove the vaccine cap and empty the large syringe. Now repeat the process with the liquid of unknown molecular weight.

The advantages of this method are that irregularities of temperature cancel out and that calculation is very simple.

From Avogadro's hypothesis we have:

$$\frac{\text{weight of 60 cm}^3 \text{ of } A}{\text{weight of 60 cm}^3 \text{ of } B} = \frac{\text{molecular weight of } A}{\text{molecular weight of } B}$$

Results

Accuracy within 3% is possible with this method provided that the boiling-point of the liquid is at least 20 °C below the temperature of the oven (i.e. not more than 100 °C if a 100-watt bulb is used).

5.5 The molecular weight of gases by rates of effusion

Fig. 5.5A

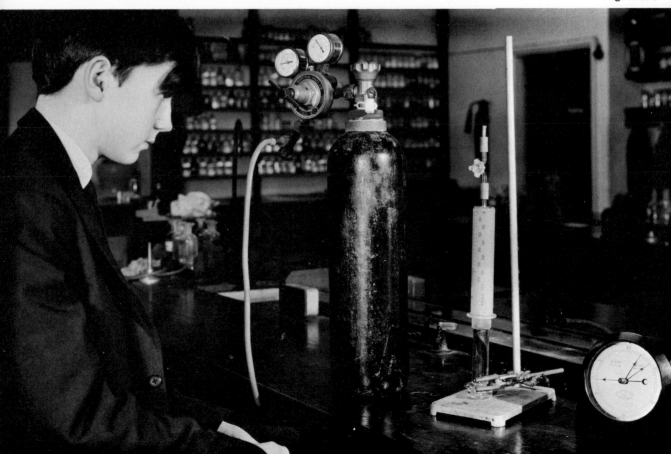

Introduction

Graham's Law states that the rate of diffusion of a gas is inversely proportional to the square root of its molecular weight (at constant temperature and pressure). A similar relationship holds for rates of *effusion* with the advantage that the relationship holds for a wider range of molecular weight than Graham found with diffusion. It will be seen that in this experiment the effusion is not simple, but the relationship still holds remarkably well.

The experiment may be treated in an interesting investigational way. Ask the pupils to find a relationship between time for effusion and molecular weight for a number of gases. Let them plot the result. Brighter ones may see that the curve they get in by a simple plot of molecular weight against time can be made more meaningful by plotting the logarithm of each value. Others may need to be told to do this. The graph obtained from the log plot is a straight line with a gradient of 2. This shows that the molecular weight is proportional to the square of the time. Time for a given volume to effuse is recommended rather than rate, for simplicity and so that those who have read ahead in their textbooks are less likely to identify the relationship with Graham's Law. Finally, you can give the pupils an unknown gas and ask them to find its molecular weight by measuring its diffusion time and using the graph to determine its molecular weight.

Apparatus

1   100 cm³ glass syringe
1   glass tube with pinhole (see below)
1   clamp, boss and stand
1   stop-clock
1   three-way stopcock (optional)

Chemicals

As many gases as are available.
Examples: hydrogen, ammonia, carbon monoxide, nitrogen, oxygen, hydrogen sulphide, carbon dioxide, butane

Making the pinhole

Take a piece of glass tubing about 2 cm long and 6 or 7 mm diameter and stick a small piece of aluminium foil over the end with Araldite. When it is dry make a small hole in the centre of the aluminium foil using the point of a fine needle.

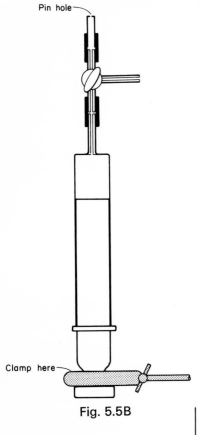

Pin hole

Clamp here

Fig. 5.5B

Procedure

Fix the syringe in the vertical position with the pinhole tube attached, as shown in Fig. 5.5B.

Flush out the syringe two or three times before filling it with 100 cm³ of the gas. Allow the gas to effuse out of the pinhole and take the time of the plunger to go from the 80 to the 20 cm³ mark. Repeat the experiment two or three times with each gas and take an average result. It is most important that the syringe should be completely free running. Any speck of dirt between the plunger and the barrel will slow down or jam the movement of the plunger and ruin the results.

The experiment is rather easier to handle if a three-way stopcock is placed between the syringe and the pinhole tube. The stopcock can be used to fill the syringe without removing the pinhole tube. It is also possible to keep it closed while the apparatus is set up and open it just at the beginning of the experiment. It is important that the stopcock is not too bulky and that the gas has free passage through it.

Results

If the syringe is truly free running good results can be obtained. Fig. 5.5C shows some results obtained at Westminster School.

Fig. 5.5C

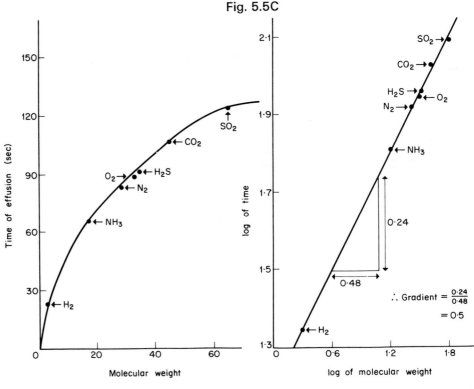

Chapter Six

Allotropy

6.1 The allotropy of carbon

Introduction

There are not many simple ways of showing that allotropes are in fact different forms of the same element. This experiment shows that a given weight of charcoal, graphite or diamond (industrial) produces the same volume of carbon dioxide when burnt in oxygen. This is evidence, if not proof, that they are allotropes of carbon. Unfortunately the form of these allotropes usually available in schools are not pure and account has to be taken of the ash remaining. Nevertheless the experiments are well worth performing. Special care must be taken in this case to see that the whole apparatus is gastight.

Apparatus

2   100 cm³ glass syringes
3   three-way stopcocks
1 (or 3 to save time) silica tube, 2 glass rods each 4 cm long to fit loosely inside the tube
1   Orsat pipette
1   manometer filled with light oil
2   syringe clamps, with bosses and stands with further clamps and stands to hold the Orsat pipette, etc., or syringe bench

Chemicals

Charcoal (small lumps – degased)
Graphite (small lumps)
Industrial diamonds (optional!). Artificial diamonds are available reasonably cheaply. Only 0·012 g are needed for this experiment
Oxygen cylinder (or other source of oxygen)
5 M potassium hydroxide for the Orsat pipette
Glass-wool

Procedure

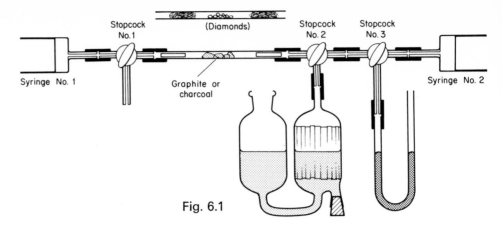

Fig. 6.1

Set up the apparatus as shown in Fig. 6.1. Syringe No. 2 should be greased. If the experiment is to be demonstrated, it saves time to prepare three silica tubes with the three carbon allotropes first. The graphite and charcoal are in the form of a small lump or lumps and are kept in place by glass rods. The diamonds are usually very small and are best kept in position in the middle of the silica tube with plugs of glass-wool. It is vital that no diamond particle should be allowed to get into a syringe. Weigh the silica tubes empty and with the carbon in them. About 1 millimole (0·012 g) will be enough. Weigh to three (or preferably four) decimal places. Note that the charcoal *must* be completely dry. It can be dried by heating it in a test-tube or by leaving it overnight in a desiccator. The charcoal must be of the degased variety. If gases are given off when the charcoal is heated the results will be inaccurate.

Graphite and diamond burn best in pure oxygen, but the charcoal should only be burnt in a 50/50 mixture of air and oxygen. Small explosions can occur if charcoal is heated in pure oxygen. Otherwise the procedure is similar for each allotrope.

Fill syringe No. 1 with 100 cm³ of oxygen (or a mixture of 50 cm³ of oxygen, 50 cm³ of air for the charcoal experiment). Make sure that the whole apparatus is gas-tight. Turn the three-way stopcocks so that syringe No. 1 is connected with syringe No. 2 but the manometer and Orsat pipette are not connected. Heat the carbon in the silica tube (very strongly in the case of diamonds) with a Bunsen burner. Pass the oxygen slowly backwards and forwards between the two syringes until all the carbon is burnt. A small amount of ash may remain with the charcoal and graphite. Artificial diamonds always leave a residue of the metal oxide mixture round which they were formed. Cool the silica tube with a damp cloth. When the apparatus has reached room temperature, push the gases into syringe No. 1 and note the volume. It should be unchanged as for each molecule of oxygen used up a molecule of carbon dioxide has been produced. This is a good example of a Gay-Lussac experiment. The pupils are sometimes surprised by the fact that the carbon has disappeared but no change in volume has occurred. If there is a decrease in volume there is probably a leak in the apparatus. If there is an increase some gases were absorbed on the charcoal.

Now transfer the gases to syringe No. 2 (the greased syringe) making sure that they are at atmospheric pressure. Turn stopcock No. 2 so that the Orsat pipette is connected to syringe No. 2. Pump the gases in and out of the Orsat pipette until no further absorption takes place. Take care not to allow any of the gases to bubble right through the Orsat pipette and out into the atmosphere. Now turn stopcock No. 2 back to its original position and push the gases once or twice backwards and forwards between the two syringes. This is to pick up any residual carbon dioxide which may be in the silica tube. Repeat the absorption in the Orsat pipette. Finally, turn stopcock No. 3 so that the manometer is connected with the rest of the apparatus and use it to ensure that the gases in syringe No. 2 are at atmospheric pressure. The decrease in volume gives the volume of carbon dioxide produced. To account for any ash which may remain, weigh the silica tube and its contents after the experiment. The difference in weights before and after the experiment gives the weight of carbon burnt.

Results

Results should be accurate within better than 5%. If results are inaccurate it is almost certainly because the apparatus is not gas-tight.

6.2   The allotropy of sulphur

Introduction

This is a very similar experiment to 6.1. The only difference is that as the sulphur is inclined to vaporise a special combustion tube is needed. The experiment can also be used to determine the formula of sulphur dioxide.

Apparatus

2   100 cm³ glass syringes
3   three-way stopcocks
1   hard glass tube, about 12 cm long with a bulb about 2 cm diameter blown 4 cm from one end
1   Orsat pipette
1   manometer filled with light oil
2   syringe clamps, with bosses and stands, or syringe bench

Chemicals

Monoclinic and rhombic sulphur
Oxygen cylinder, or other source of oxygen
5 M potassium hydroxide for the Orsat pipette (sulphur dioxide is
  so soluble that water will do)

Procedure

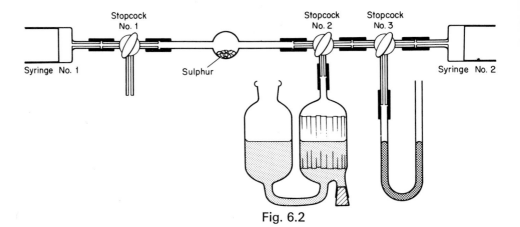

Fig. 6.2

Follow a similar procedure to that given for experiment 6.1. Pure
oxygen is used for burning the sulphur. The sulphur in both cases
should be crystalline (or solid chips) but not in powdered form. Weigh
the hard glass tube empty and with about 2 millimoles (0·064 g) of
sulphur in the bulb. When the apparatus is set up (note that the bulb
is nearest syringe No. 1) fill syringe No. 1 with 100 cm³ of oxygen.
Test the apparatus to see that it is gas-tight. Turn the stopcocks so
that syringe No. 1 is connected through the hard glass tube to syringe
No. 2. Heat the silica tube strongly with a Bunsen burner flame beside
the bulb on the side nearest syringe No. 2. Then heat the bulb of the
hard glass tube until the sulphur begins to melt. Then pass the oxygen
very slowly over it. When all the sulphur has burnt cool the apparatus
down and proceed as in experiment 6.1.

Results

Good results will be obtained provided that the sulphur burns without vaporising and condensing unburnt in other parts of the silica
tube.

Chapter Seven

The Physical Properties of Gases

7.1 Boyle's Law

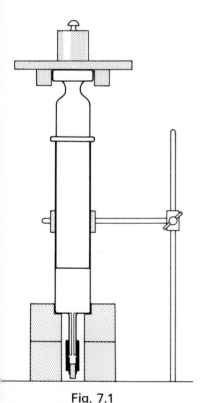

Fig. 7.1

Introduction

The syringe provides a simple way of measuring the effect of pressure on the volume of a gas. The experiment is best performed with a special holder for the syringe and a 'table' for the weights. These can easily be made in the laboratory.

Apparatus

1   100 cm³ glass syringe (greased)
Glass rod and rubber tube to close syringe
Holder for syringe and 'table' for weights (see Fig. 7.1)

Procedure

Before starting the experiment note atmospheric pressure, measure the diameter of the syringe plunger accurately and weigh the plunger and the 'table' for the weights. Note that the syringe should be greased. The experiment can be performed by adding weights to the table and by hanging weights from it, as can be seen from the diagram. Draw in about 60 cm³ of air with the syringe. Add the weights 1 kg at a time. Repeat the experiment with each weight three times and take the average.

Results

Results should be correct to within the accuracy of the syringe.

7.2   Charles' Law

Introduction

The relationship between the temperature of a given mass of gas and its volume can be measured very simply by means of a syringe. The means of heating can vary. A water bath has limited application, but it is excellent within its limits. Electric furnaces give a much wider range of temperatures.

Apparatus

1   100 cm³ glass syringe
1   three-way stopcock
1   syringe furnace (or water bath)
1   thermometer
Vaccine cap, or stopcock to close syringe

Procedure

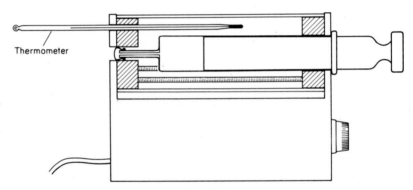

Thermometer

Fig. 7.2

Draw in 50 cm³ of air into the syringe and place the syringe in the furnace as shown in the diagram. Raise the temperature of the furnace from room temperature at 10 or 20 °C intervals. Allow the furnace and syringe at least five minutes to equilibrate at each temperature. At higher temperatures the barrel and plunger tend to expand at different rates. It is advisable therefore to grease the syringe lightly. Care must be taken to 'feel' that the syringe is at atmospheric pressure.

The results of this experiment may be compared with different gases. See Chapter Four for examples of gaseous dissociation at high temperatures.

Results

The difficulty here is that syringes tend to leak at higher temperatures; especially if they have not time to equilibrate. If the experiment is taken slowly an accuracy within $\pm$ 5% can be expected.

7.3   The solubility of gases : very soluble gases

Introduction

The syringe affords a very simple way of determining the solubility, by volume, of a soluble gas. This experiment can easily be performed by pupils using glass or plastic syringes.

Apparatus

1   100 cm³ glass syringe or 50 cm³ plastic syringe
1   vaccine cap
1   1 cm³ syringe (plastic or glass)
1   20 cm³ plastic syringe
1   syringe needle

Chemicals

Water soluble gases such as: ammonia, carbon dioxide, chlorine, hydrogen chloride, hydrogen sulphide, nitrogen dioxide (but note the association to dinitrogen tetroxide), sulphur dioxide.

Procedure

Fill the 100 cm³ glass syringe with 60 cm³ of the gas (or the 50 cm³ plastic syringe with 30 cm³ of the gas) and seal the syringe by placing a vaccine cap over the nozzle. Inject a known volume of water into the 100 cm³ syringe via the vaccine cap. In each case the volume of water injected will be different. The most soluble gases such as hydrogen chloride and ammonia will only need 1 cm³ of water for 90 cm³ of the gas. Carbon dioxide needs 40 cm³ for 60 cm³ of the gas. The solubilities of gases in water at room temperature given below may act as a guide to show how much water to inject. Shake the syringe and solution well before measuring the decrease in volume of the gas. The volume of solution is of course taken into account. Note room temperature and atmospheric pressure. Solubilities at different pressures and temperatures can be obtained by combining this experiment with experiments 7.1 and 7.2.

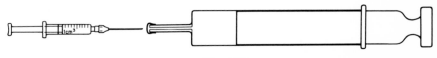

Fig. 7.3

Solubilities of gases   (from *Physical and Chemical Constants* by Kaye and Laby) cm³ dissolved in 1 cm³ of water at 760 mm pressure and 15 °C.

Ammonia 801
Argon 0·037
Carbon dioxide 1·019
Carbon monoxide 0·025
Chlorine 2·63
Helium 0·0089
Hydrogen 0·0188

Hydrogen sulphide 2·91
Hydrochloric acid 458
Nitrogen 0·0179
Dinitrogen monoxide 0·74
Nitrogen monoxide 0·051
Oxygen 0·034
Sulphur dioxide 47·3

7.4   The solubility of gases: sparingly soluble gases

Introduction

This experiment is suitable for fairly soluble gases like carbon dioxide (see experiment 7.3) and for sparingly soluble gases like oxygen and nitrogen. The technique is relatively simple and the experiment takes only a short time to perform. It is suitable both for demonstration and for the pupils to do themselves. The experiment was devised by Mr C. V. Platts.

Apparatus

1   100 cm³ glass syringe
1   three-way stopcock
1   1 litre capacity separating funnel (for carbon dioxide 250 cm³ capacity is sufficient)
1   bung to fit the separating funnel with hole to fit the three-way stopcock
1   measuring cylinder (largest available up to 1 litre)
1   syringe clamp, with boss and stand
1   standard clamp, with boss and stand

Chemicals

Slightly or sparingly soluble gases such as: carbon dioxide, dinitrogen monoxide, carbon monoxide, nitrogen, oxygen, etc.
Boiled distilled water

Procedure

Before beginning the experiment, boil several litres of distilled water to get rid of dissolved gases and allow the water to cool in a flask sealed from the air. The quantity of water required will depend, of

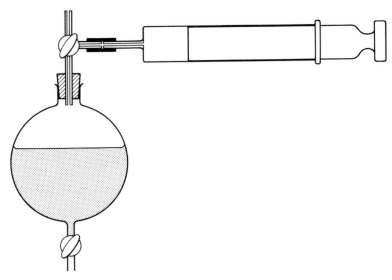

Fig. 7.4

course, on the number of experiments to be performed. Each experiment with a sparingly soluble gas needs 1 litre of water. The more soluble gases need only 250 cm³.

Fill the separating funnel with degased water. Attach the bung, syringe and three-way stopcock. Attach the cylinder of gas (or apparatus for producing the gas) to the open end of the tap and turn the stopcock so that the syringe, but not the separating flask, is connected. Flush out and fill the syringe with 100 cm³ of the gas. Now turn the stopcock so that the cylinder and the separating funnel are connected, but the syringe cut off. Open the tap on the separating funnel and draw in the gas from the cylinder at the same time allowing half the water to pour out of the separating funnel. Turn off the separating funnel tap, disconnect the cylinder and turn the stopcock so that the syringe and the separating funnel are connected to each other but not to the atmosphere. Shake the funnel vigorously until no further decrease in volume is shown on the syringe. Note the decrease in volume, which is, of course, equal to the volume of gas which has dissolved in the water. Pour the solution out of the measuring cylinder to determine its exact volume. Find atmospheric pressure and note room temperature.

Results

This experiment gives results which are within the accuracy of the syringe. To decide whether to use a 250 cm³ or 1 litre separating funnel consult the list of solubilities of gases given after experiment 7.3.

7.5 The effect of changes of pressure on the temperature of a gas

Introduction

There are several experiments which show that the temperature of a gas rises when its pressure is increased and falls when its pressure is decreased, but this one shows it rather neatly and directly.

Apparatus

1   100 cm³ glass syringe
1   thermistor fixed into a tube as shown in the diagram
1   Avometer or alternative means of recording the change of resistance of the thermistor

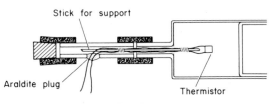

Fig. 7.5

Procedure

Making the thermistor holder:

The object is to place a small thermistor inside the syringe about 1 cm from the nozzle outlet. The thermistor must be small enough to be inserted through the nozzle. If this is not so an alternative method to the one described here must be used. Blow a small hole (about 5 mm in diameter) in the side of a length of soda-glass tube of about 8 mm outside diameter. Seal one end of the tube. Fix the thermistor to one end of a splint of wood (a cocktail stick will do) so that it can be pushed through the nozzle of the syringe without the two lead wires shorting. Seal the thermistor into the tube by means of a blob of Araldite.

Draw in some 60 cm³ of air (or other gases) into the syringe and then insert the thermistor through the nozzle. Use a short length of rubber tubing to attach the glass tube to the nozzle of the syringe. The gas in the syringe is now sealed from the atmosphere. Connect the two protruding wires from the thermistor to an Avometer set to measure resistance; or to a thermistor circuit. Holding the tube with one hand to prevent it being blown off the nozzle compress the gas in the syringe by pushing in the plunger sharply. The thermistor will record a rise in temperature, which is quickly dissipated. Similarly drawing out the plunger sharply will show a lowering of temperature. This experiment is useful qualitatively rather than quantitatively but it is interesting to note the behaviour of different gases.

Index